THE WEIL CONJECTURES

THE WEIL
CONJECTURES

ON MATH

AND THE PURSUIT OF

THE UNKNOWN

KAREN OLSSON

FARRAR, STRAUS AND GIROUX   NEW YORK

Farrar, Straus and Giroux
120 Broadway, New York 10271

Printed in the United States of America
First edition, 2019

Library of Congress Cataloging-in-Publication Data
Names: Olsson, Karen, [date] author.
Title: The Weil conjectures : on math and the pursuit of the unknown / Karen Olsson.
Description: First edition. | New York : Farrar, Straus and Giroux, 2019.
Identifiers: LCCN 2018052629 | ISBN 9780374287610 (hardcover)
Subjects: LCSH: Weil, André, 1906–1998. | Mathematicians—France—Biography. |
    Weil, Simone, 1909–1943.
Classification: LCC QA29.W455 O57 2019 | DDC 510.92 [B]—dc23
LC record available at https://lccn.loc.gov/2018052629

Designed by Abby Kagan

Our books may be purchased in bulk for promotional, educational, or business
use. Please contact your local bookseller or the Macmillan Corporate and
Premium Sales Department at 1-800-221-7945, extension 5442, or by e-mail
at MacmillanSpecialMarkets@macmillan.com.

www.fsgbooks.com
www.twitter.com/fsgbooks • www.facebook.com/fsgbooks

10 9 8 7 6 5 4 3 2 1

For Andrew

PART ONE

SMALL FOR HER AGE, SHE TAKES UP MAYBE a third of the operating table, from her folded-over socks to the crown of her head. Every so often a nurse places a towel soaked in chloroform over her nose and mouth. While the doctor cuts into her abdomen she babbles, sings Christmas songs, recites names from ancient legends. She seems not so much asleep as bewitched.

It's 1912. Simone is three.

With a snip and a pinch Dr. Goldmann fishes out her swollen appendix. "We are not interested in luxury," the tiny patient announces, and the forceps nearly fall from his hands. As though she's been possessed by some restless soul—the things she says! Afterward he tells her mother that she is too extraordinary to go on living.

She recuperates in a drafty room in the hospital annex. Although her mother urges her to lie still under a pile of blankets, Simone makes a game of kicking off the covers and trying to wriggle out of bed. As a young woman, Selma had wanted to study medicine, but her father, Simone's grandfather, forbade it, and instead she married a doctor, drew a magic circle around her family, and lavished herself upon them. Now Selma brings a basin to the bed so that Simone can wash her hands for the fourth time. Now she tries to pin down her daughter by telling her a story called "Marie in Gold and Marie in Tar."

Once there was a girl named Marie who was sent by her stepmother into the forest to look for food. In the middle of the forest she came to a house, and there she heard a voice asking whether she would like to enter through a door of gold or a door of tar.

"Tar is good enough for me," she said, and all at once she was showered with gold pieces. She brought them home to her stepmother, who promptly sent her own daughter, also named Marie, into the forest. This second Marie arrived at the same house and was asked the same question. She, however, picked the door of gold. As soon as she walked through it, she was deluged with tar.

I am left wondering whether the first Marie ever found any food.

Whether the second Marie died of asphyxiation.

At any rate, Simone would later say that the story had a profound influence on her. All her life, it seems, she went looking for tar doors.

Though the doctor couldn't have meant it literally when he judged her unfit for this life, he would be proved right. She was extraordinary, and at thirty-four she died a very strange death—an extraordinary death, the end product of her extraordinary manner of thinking and living.

Her older brother, André, is also extraordinary. A few years after Simone's appendectomy, when he's about nine, he discovers an algebra book, its own kind of door. He races right through, enters a house full of equations, and begins to tinker with them, rearranging terms, expanding here, factoring there. Sometimes when he finishes a page of calculations he'll hold it up above his face and admire the sheer density of what he's written, everything in alignment. The superscripts, the equals signs, the variables: x after x after x.

This worries his mother. "We can't help but be a trifle annoyed and anxious, my husband and I, at the absorbing passion André shows for algebra," Selma writes in a letter. "Some way or other he got hold of a book . . . and he is so happy that he has given up all play and spends hours immersed in his calculations."

A month earlier, he'd been just as obsessed with croquet.

One day, coming down the stairs, he stumbles and falls. As he sits on the floor, clutching his knee, Simone, aged six, rushes to find his algebra book and then brings it to him, because she's sure it's the thing that will comfort him most.

His parents try taking away his paper and pencil, so that he might spend more time outside, but soon they find him using a pebble to scratch equations into the sidewalk.

———

A head for numbers.

Episodes of graphomania turn up in biographies of other mathematicians; take for instance Archimedes, who, it is said, would rake the ashes out of a fire and draw shapes in them. Who after bathing and oiling himself would trace diagrams upon his skin with a fingernail.

Or Karl Weierstrass, who in the 1840s and '50s taught school by day and reimagined the field of mathematical analysis by night. Give him a pencil, said one of his sisters, and he might start to scribble his mathematics on any surface—on wallpaper, on a shirt cuff.

Weierstrass would later write of the "infinite emptiness and boredom" of his years as a schoolteacher, before he achieved renown as a mathematician just shy of forty.

About his success he would say, "Everything in life comes too late."

The scrape of the rock across the pavement as the boy works out his calculations. His delight as terms fall away and roots reveal themselves, as he begins to hang formulas and number relationships around the new rooms in his mind. He takes shelter there while autos throttle past and cannons fire in the distance—the country is at war. The family has followed Dr. Weil to Neufchâteau, where he attends to soldiers wounded in battle and to victims of a typhoid epidemic.

For the sick patients the preferred treatment is to plunge them into ice-cold baths. Most of them die.

André still plays with his sister, usually tutor to her pupil. He teaches her to read, delivers astronomy lectures on the bus. A know-it-all in short pants and a doll of a girl in a sailor dress, her hair in ringlets, egging him on with questions: oh, how they irritate the other passengers with their precocity! Something they don't yet know is that they are Jewish, since their parents have turned away from religion and never discuss it, but there's bound to be at least one old bigot on the bus who presumes by their appearance. Who huffs, *I can't stand listening to children parrot things they don't understand.*

Simone and André memorize long sections of verse by Corneille and Racine, and they recite them in turn, staring bug-eyed at each other. It's a contest: although they smirk as they call out the lines, every time one of them misses a word or mangles a phrase, the other delivers a hard slap to the face.

They look alike, an older boy version and a younger girl version of the same child, with abundant black hair and eyes full of dark intensity, hungry to know everything. Churning brains appended to small, floppy bodies. They'll shout a phrase in ancient Greek and take off running. Or fall so deep into their novels as to forget lunch. Or cackle loudly at some joke that only they find funny, tumbling into each other, their laughter a pulse that bounces back and forth between them. Their mouths gaping, their noses practically touching: ha ha ha ha ha!

Or: from another room Selma will hear odd shuffling sounds, and when she comes in to investigate she finds them pinching, kicking, grabbing each other by the hair. Their faces have gone white with rage.

One winter, André decides that he is done with knee socks, he wants to tough it out bare-shinned, and, naturally, Simone follows his lead. Over their mother's protests they go sockless; worse, they'll board a bus with her and shiver theatrically and chatter their teeth and complain that their parents refuse to buy them socks. One day another passenger hurries out of the bus after them, points at Selma, and denounces her: "You wretch!"

Children's games, but in adulthood, too, Simone would be eager to deprive herself, to go without heat, food, sleep. And an inner voice would lash at her, *You wretch!*

"La Trollesse" is her childhood nickname within the family. The (feminine) troll, as though she were not quite a girl, yet not *not* a girl. And not quite of this world, though not belonging to any other: a lumbering creature from a fairy tale.

Once there were a brother and sister who devoted themselves to the search for truth. A brother who spent his long life solving problems. A sister who died before she could solve the problem of life.

2. IT WAS A CONVERSATION WITH HIS printer that led René Descartes to choose x to represent unknowns in his 1637 treatise *La Géométrie*, or so the story goes. Running out of letters, the printer offered Descartes the choice of x, y, or z to indicate unknown quantities in equations. When Descartes replied that it didn't matter to him which one they used, the printer selected x, because x was used less frequently in French than y and z. In other words, a practical suggestion by a seventeenth-century typesetter lies behind all the x's in algebra, and maybe some other x's too. One way or another, x has come to stand for what we don't know, what we're seeking, for sex shops and invisible rays and the marked spots where treasure lies hidden.

Was the choice strictly pragmatic, I wonder, or was there always something erotic about x?

To compress the unknown into a single symbol was very powerful, for it made algebraic manipulation easier and more legible, and after Descartes it was widely adopted. Powerful in a broader sense, too, this naming the unknown: x, neither alive nor dead, took on an existence of its own, outside of time, worming its way into an infinitude of equations, of propositions. The unknown as a *thing*.

What is my unknown? My *x*?

"An insect tries to escape through the windowpane, tries the same again and again, and does not try the next window which is open and through which it came into the room," writes the mathematician George Pólya in his book *How to Solve It*. "A man is able, or at least should be able, to act more intelligently. Human superiority consists in going around an obstacle that cannot be overcome directly . . ."

But I feel for the insect, because I'm a person who tends to be drawn to the glass, the banging—banging my head against a see-through wall, banging instead of solving, a way of forever putting off the solution. Then again I think there's an intermediate condition, a mode that is more than banging and less than solving. It's this in-between state I like, as I buzz my way across the window glass, not quite bent on escape.

Simone on the verge of adolescence: a slip of a girl half hidden by boxy clothes, glasses, that curly mass of hair. Skinny legs and small, uncouth hands, like buds reluctant to bloom. Her mother favors boys—the "forthrightness" of boys, she once explained in a letter, as opposed to the "simpering" of girls—and for a while Simone wishes to be treated as a boy.

Her family calls her Simon; she signs letters to her parents "your respectful son."

By the age of thirteen or fourteen, it becomes much harder for a girl to play at being a boy, and that's not the only illusion that won't hold up. Simone enters into a period of what she'll later call "bottomless despair." A sinkhole of self-doubt and shame. She decides she isn't especially smart, not smart like her brother is, and sees no point in living if she's merely a normal person.

She thinks about killing herself. "I didn't mind having no visible successes, but what did grieve me was the idea of being excluded from that transcendent kingdom to which only the truly great have access and wherein truth abides," she will later recall. "I preferred to die rather than live without that truth."

Why should her brother be admitted to the kingdom and not she? What else was there?

Then, after weeks, months, of self-loathing, she discovers a way forward. A tar door. *When one hungers for bread, one does not receive stones.* Anyone who is sufficiently patient may achieve a kind of transcendence, provided that he "longs for truth and perpetually concentrates all his attention upon its attainment," she'll write. She arrives at an idea of strenuous faith, a discipline of attention. It's a crucial epiphany for her: the only way she can rescue herself, that is to say the only way she can (on her terms) lead a life that is not worthless, is to devote herself wholly, with every ounce of her energy, to the truth—an impossible goal, really, but she would stay dedicated to it.

Her brother, by this time, is a student at the renowned École Normale Supérieure, a temple of learning where he gorges himself on mathematics and also on languages, among them Sanskrit. He's rarely home, and when he is home he has no time to explain anything to her.

Her relations with her lycée classmates are not bad, but they're not especially good either, and while she endures their pranks and shares meals with them, she invents a secret friend. This friend is, curiously, distant and hidden, a friend who she hopes will be revealed to her one day. She has made up a friend who won't keep her company.

I know I wasn't the only high-school girl to check *The Simone Weil Reader* out of the library. Saint Simone, herself an imaginary friend to who knows how many lonely teenagers of a certain era. In her own way, a severe and elusive one.

Did I actually read *The Simone Weil Reader*? Or did I just flip through it, lying on my bedroom carpet, depressed and restless, listening to cassette tapes? As a gawky girl, circa 1989, I was less curious about her writing than I was about Simone herself, the petite French ascetic with cool hair and wire-frame glasses, the political activist / intellectual / mystic who died young. Casting about for role models, I chose the most outlandish ones, women who'd lived lives I would never lead, who'd suffered in ways I never would, uncompromising and bold and pure while I was none of those things.

Maybe Simone and I have some unfinished business, that is to say, I'm moved by André Weil's story in part because he was the brother of Simone Weil, the luminous mystery behind a book I failed to read. But I could also say that I'm drawn back to Simone Weil because she was the sister of André Weil, one of the great mathematicians of the twentieth century—math representing, for me, another piece of unfinished business.

Beginning in 1897, the mathematician Felix Hausdorff published literary essays under the nom de plume Paul Mongré, and in one of them he laments that two genders are insufficient. He wishes for a third: "Everything is thus, but must it be thus? . . . Could there not be a crystal space, where one could see around the corner and sense one's way into another I? . . . Or why not three genders . . . Men, Middlers, Mothers . . ."

Never mind the dubious setup—that is, "men" at one pole and "mothers" at the other—I like his notion of "middlers." I think of myself as a middler, or used to think of myself that way before I became a mother. Once I was a tall boyish girl who liked math—which, much as it was a field dominated by men, struck me even more so as a field dominated by middlers, by which I mean that in the aggregate its people seemed more androgynous than the general population.

Mongré = *mon gré*, French for "my taste" or "my liking." Under that name Hausdorff also published poetry, a play, a book of aphorisms, and a book-length philosophical essay titled *Chaos in Cosmic Selection*.

Simone enters the École Normale Supérieure herself in 1928 to study philosophy, one of a handful of women enrolled there. She smokes constantly, loves to stay up all night talking and arguing in a café, though what she'd prefer to be doing (or at least thinks she'd prefer to be doing) is manual labor. She wants to work on a farm. She believes it would've been better to be born poor. Once, while walking with a fellow philosophy student named Camille Marcoux, she passes a wine market and decides to apply at once for a job putting corks into wine bottles. Marcoux has to steer her away.

Like Simone, Marcoux is interested in mathematics, and he lends her a geometry textbook by Jacques Hadamard, a beloved professor at the École Normale and an outstanding mathematician—who, as it happens, would become the supervisor of André's doctoral thesis on Diophantine equations. When she returns the book, it has been all but torn apart, with some pages marked up and others ripped out. Hadamard, she informs Marcoux, has committed crimes against geometry.

(What these crimes might've been, I don't know.)

Then there's a vacation on the Normandy coast in 1931. Simone is captivated by the fishermen there and begs them to let her join one of their crews. They all reject her, until one of them, Marcel Lecarpentier, sees her "running along the shore like a madwoman," as he'll later recall, then "going into the sea with her wide skirts," and turns his boat around. He allows her onboard to work. When a bad storm hits, she refuses to be tied down.

"I'm ready to die," she declares. Her voice is peculiar, a

low monotone and yet full of fervor. "I've always done my duty."

My own earliest math-related memory dates to around second or third grade, when "function machines" appeared on worksheets. These were cartoon machines, tall and narrow and amiable, that turned numbers into other numbers. Each machine contained a two-column grid, partially filled with numerical inputs and outputs. On top of the grid was a space for a rule, like "Multiply by five," in which case the number in the first column times five would give you the number in the second column. The task was to supply the blank parts of the grid, to write the machine's output for a given input, or vice versa. Or sometimes you'd have to guess the rule from the numbers given. All of which is just to say that I remember being pleased by the machines; I liked the idea of a multiply-by-five contraption. It was an early step back from numbers themselves, a shift of emphasis toward the dynamics of it all, from the stolid nouns to the freewheeling verbs that connected them.

And a little green shoot in my mind: there must have been something tantalizing to me about abstraction itself, even if I couldn't have said as much at the time. I remember in that same year discovering a pleasure in writing that was in no small part physical, delighting in the way I could move a pencil across my notebook and fill line after line, entire pages! Of course I didn't recognize how I'd begun to reposition myself, how ready I was to disappear into a piece of paper—how the representation of a thing could seem more alluring than the thing itself.

————

Hausdorff rose to fame (mathematical fame, at least) for his work on set theory, which helped lay the groundwork for modern topology. He was a professor in Leipzig, then Bonn, then the Baltic city of Greifswald, and then Bonn again.

Though Jewish by birth he had assimilated; his wife, Charlotte, the daughter of a Jewish doctor, had converted to Lutheranism, and they baptized their daughter. Believing that they could keep their heads down and get by unnoticed, they reacted too late to the rise of the Nazis. On November 9, 1938, the day after Felix's seventieth birthday, came the pogrom known as Kristallnacht. A mob gathered outside his house.

"There he is, the head rabbi," they shouted. "Just watch out. We are going to send you to Madagascar, where you can teach mathematics to the apes."

As I understand it, André and Simone Weil's last name is pronounced something like "vay" but every time I read it I hear the word *wail*. André wail. Simone wail.

After Kristallnacht, Hausdorff searched for a means to emigrate to America. He never found one. In 1942 he, Charlotte, and Charlotte's sister took lethal doses of poison in order to avoid the camps.

"I am sorry to cause you yet more effort beyond death," Hausdorff wrote in a farewell letter to Hans Wollstein, who was his friend and estate lawyer, and who himself later died at Auschwitz. "Forgive us our desertion! We wish you and all our friends to experience better times."

A masochistic student, Simone spreads out her books on the floor of her unheated apartment and shambles on her knees from one text to another, the winter wind blowing through the open windows as she turns the pages.

She crawls about the room, then leans over Descartes like an animal drinking, her vertebrae bulging through the back of her shirt. Then it's on to Kant, to Chardin. There is damp laundry hanging on lines strung between the walls, stiffening in the cold, a banner that she forgets about when she pushes herself up and stands on tingling feet. For a moment her face is shrouded in her underwear.

Or she bows over a geometry book at a quay of the Seine where large blocks of stone are unloaded from boats, kneeling on the ground here too, in love with stone, with everything that's hard.

"I studied mathematics, which is the madness of reason," announces the narrator of Clarice Lispector's *Água Viva*.

While in college André learns enough Sanskrit to read the Bhagavad Gita, with the help of an English translation contained in an anthology called *Sacred Books of the East*. He is smitten with the poem, and it becomes his guide, as much of a faith as he will find in his life. Later on it will also help him to intuit something of the way his sister thinks.

His copy of the Gita is a small volume covered in red velvet, the pages coarse and pulpy. The black script bleeds in places. He mutters lines to himself, undulations of syllables.

O Sanjaya, what did my sons desirous of battle and the sons of Pandu do after assembling at the holy plain of Kuruksetra?

He meets—in Paris, at the home of a French scholar of Indian studies—the minister of education for the state of Hyderabad, a tall, laughing man who says he'll soon be appointing new professors to the Aligarh Muslim University, not far from Delhi. The minister is to become president of the university and wants to establish a chair of French civilization. André, who yearns to go to India on any terms, volunteers for the job. Later that fall he receives a cable: IMPOSSIBLE CREATE CHAIR FRENCH CIVILIZATION. MATHEMATICS CHAIR OPEN. CABLE REPLY.

Long ago, in the prehistory of civilization, the human mind was crude, a basic animal mind—or so conjectured Simone in "Science and Perception in Descartes," a long essay she wrote during her final year of the École Normale. This strange paper contains in embryo some of the questions and themes that would obsess her throughout her life. Hardly a conventional academic work, it starts off with a speculative tale of how science came to be: In some ancient era, she imagines, people had no access to broader concepts and abstractions, yet they had an inkling that other forms of thought were possible, and so they invested priests and kings with power, because they believed them capable of this higher knowledge. And then along came Thales, an early Greek mathematician. His development of geometry was "history's greatest moment," she writes—(!!!)—a revolution that overthrew the absolute authority of the priests.

But to what end? she asks. Did that revolution merely replace the false rule of priests and kings with a (truer, yet still unjust) rule of mathematicians and scientists? Or did it bring equality, by revealing that the purest thought is also ordinary thought, that we might all live by the light of our own minds?

Recently I sought out this essay, and I'd hardly begun reading before it became obvious to me why I never made much headway into *The Simone Weil Reader*. Her prose is dense, at times baffling; I have to bushwhack my way through "Science and Perception in Descartes," clearing back each sentence, each long paragraph, without much notion, as I go along, of where I'm headed. Many of her claims are more bold than they are convincing. Surely a professional historian would reject the essay out of hand. But as I read it, I picture its author, a twenty-one-year-old philosophy student still struggling to step out of her brother's shadow, and I find even her oblique flights of theorizing more fraught, more poignant, than I would have as a younger reader, since now I take this parable to be informed by Simone's own youth. I remember the specter of her older brother's genius, her own crisis of identity, and I think *of course* she associates power with some inaccessible type of cognition.

Once she's done with her imaginary history, André's invisible pull on the essay seems all the stronger. For Simone then launches into a critique of present-day math, or as she puts it, "the absolute dominion that is exercised over science by the most abstract forms of mathematics." Math has drifted too far away from us; it has become disconnected from the real world. Eventually she'll take this up with André directly and tell him outright that contemporary mathematics, his beloved vocation, is too removed from life.

She wants to reconcile the abstract and the concrete, to make philosophers and mathematicians of us all, but how that would work is never clear. Math, as Simone sees it, ought to function as a kind of passageway between the mind and the world: "I am always double: on the one hand, a passive being who is subject to the world, and on the other, an active being who has a grip on it; geometry and physics help me to conceive how these two beings can be united, but they do not unite them." Only through action—"real action, indirect action, action that conforms to geometry"—can reason seize hold of the world.

Action that conforms to geometry? What could that be? During the year she worked on the essay she barely consulted her adviser, the philosopher Léon Brunschvicg, who in the end did not think much of "Science and Perception in Descartes." He gave it the lowest possible passing grade: a ten out of twenty.

———

But here's another line from the essay: "I must be tricky, cunning, I must hamper myself with obstacles that lead me to where I want to go."

Who could say what that line has to do with Descartes, still I think to myself, *Yes, exactly*—this is how a writer must be. All these years I've spent throwing obstacles down in front of myself, coming up with problems too twisted to solve.

Everything in life comes too late.

3.

HAVING FORGOTTEN WHATEVER I ONCE knew about complex functions Fourier series field extensions compact surfaces hyperbolic spaces random walks et cetera, I am left with memories of the Science Center at Harvard, a building with a facade like stair-stepped boxes, constructed around the time I was born, in the early seventies. I attended classes there most days and spent I don't know how many nights working on problem sets in the library or in vacant rooms. I remember tracking slush through an entry already muddied by hundreds of boot prints, coming and going, descending to the basement computer center or landing in one of the main-floor lecture halls or making my way upstairs to a classroom empty of charm or even the notion of charm. Though elsewhere literature and history were taught in stately old rooms softened by high windows and wainscoting, historically appropriate paint colors, the mossy aura of textual study accompanied by a certain whiff of wealth, the sciences had been paired with austere minimalism, that is to say white walls and black chalkboards and silver conduit pipes leading to clusters of heavy-duty switches. I sat among my fellow students (whizzes, immigrants, nerds, with all their anxious, humming brainpower) in a plastic chair, my backpack at my feet, always afraid that I was about to fall hopelessly behind but also proud that I'd so far managed to hold on to the fast horse of a difficult class. I who, as a white girl from private school, might've seemed marked for the humanities but who had wandered over there instead, as though by mistake. Not a boy, not Asian or Indian or Jewish, not from Russia or eastern Europe, not a child of scientists. It seemed as if all the other math kids belonged to one or more of those categories.

As for why I touched down in their midst: I could blame my erratic curiosity, a tendency to follow my nose no matter how many times my nose has led me astray. Or I could say that I was trying to prove myself, to no one other than myself. But after more than two decades, if anything my dalliance with math seems like just that, a past love, one I remember with nostalgia and the kind of echo feeling that adheres to the memory of an old romance. I mean, I had always liked math, but just how it came to consume me in college is a question that I produce a different answer to each time I'm asked— whether it's somebody else who's asking or whether I'm asking myself, as I still do from time to time.

An Italian ship takes André from Genoa to Bombay, a two-week voyage, and when the weather is good he strolls back and forth on the deck, reciting lines from a Sanskrit poem. To shield himself from the southern sun, he wears a cork helmet he bought in Paris, and yet, surrounded by sparkling ocean, he burns in no time. His skin peels, and over the course of two weeks he turns the red-brown color of a fox.

mandam mandam nudati pavanah
(gently gently blows the wind upon you)

From Bombay he goes by train to Delhi and then on to Aligarh, where the two men who've been sent to greet him titter at his helmet. He'll settle into an adobe house with a zoologist from Germany: no electricity or plumbing, but high ceilings and a roof terrace and furniture that he has custom-made, according to French designs. His office at the university overlooks a courtyard, which every day fills with students and empties out again, then just before sunset fills with long-tailed birds that speckle the grounds with their shit.

A few doors down is a gnarled old chemistry professor who complains about his idiotic students, about his paltry wages, and especially about the lowlife employees of the railway station who, he's sure of it, have stolen the shipment of guavas that he ordered from a faraway orchard. He claims he can smell the fruit in the station, that he has found guava seeds near the tracks.

Some of the oldest recorded mathematics comes to us from ancient Egypt, in documents such as the Rhind papyrus, which dates from around 1700 B.C. and contains eighty-five problems and their solutions. The opening line of the papyrus has sometimes been translated as "Directions for Attaining the Knowledge of All Dark Things."

It was a dark thing, perhaps, to find the volume of a truncated pyramid, as in this problem: "If you are told: A truncated pyramid of 6 for the vertical height by 4 on the base by 2 on the top. You are to square this 4, result 16. You are to double 4, result 8. You are to square 2, result 4. You are to add the 16, the 8, and the 4, result 28. You are to take a third of 6, result 2. You are to take 28 twice, result 56. See, it is 56. You will find it right."

My freshman year, during the first week of classes, when you could attend a lecture or two before committing yourself, I visited a daunting, yearlong math course meant for prospective math majors. By the time I found the room, a narrow auditorium with a sloped floor, the seats were already taken. I stood in the back and peered down at the professor, who seemed really far away, not only because he was in the front of the lecture hall and I was in the rear but because of his thick glasses, and the way he spoke in damp, guttural torrents inflected by what might've been a mild speech impediment, not to mention the very energy in the room. A geeky electricity.

CONSIDER A BALL IN N DIMENSIONS, the professor said.

Below me, a pack of heads bobbing, nodding, while he chalked a mysterious inequality up on the blackboard. A ball in *n* dimensions? I had no idea what that meant, much as I would've liked to think it referred to a fancy-dress occasion in an alternate universe. I left and took a computer science class instead.

But I turned out to be lousy at computer science—I had no patience for debugging programs—and at the beginning of my sophomore year I went back to the same math class. Which was muleheaded of me, since in the prior semester, in addition to bombing out of second-semester computer science, I had not done especially well in a multivariable calculus class meant for physics and engineering majors. I'd pretty much decided to leave all that behind and major in philosophy. But I can remember reading *Hamlet* at nineteen and understanding even that play as essentially a story of a fellow adolescent who, like me, was indecisive. And there were other factors, forces that drew me back. Before that second semester came to its sorry end, I'd signed up for a math summer program, and there I learned that the math that professional mathematicians do has a different tenor from multivariable calculus for physics majors, more abstract and more rigorous. That summer I also began an epistolary romance via a platform called Pine, an early form of e-mail used mostly by science people (and all too perfectly named, for a facilitator of epistolary romance), which would lead to an actual romance with someone studying physics and math. In other words, I found some social support for the whole endeavor. In other words, I fell in love. And then there was the very fact that I'd felt defeated—I wanted to prove those other classes wrong.

By that time I had overcome my fear of the ball in n dimensions, and I knew that the course split into two classes after the first exam: one faster-paced and directed at the sort of kids who seemed to have all met one another already at International Math Olympiad competitions; the other, while still challenging, accessible to more ordinary people who happened to like math. I took that other class, which as it turned out was the best class I ever took in anything.

———

Though André is a twenty-three-year-old foreigner in his first proper job, he's nonetheless been made chairman of the department. Right away he is saddled with a complicated subtraction problem: he must produce a report on the staff of his department and in effect choose which of his three colleagues should be fired. None had made a good impression. "Pathetic characters," he'll call them in his memoir, "devoid of merit."

One of them short and obsequious. One of them with a long beard he dyes red, known for his willingness to help students. One who claims to be studying a copy of an ancient Arabic manuscript, yet no one besides the man himself has ever laid eyes on the putative document. If André knew where to find decent replacements, he would happily fire them all.

The best class I ever took in anything, not just because I was entranced by the math itself but because we were encouraged to work on the difficult weekly assignments in groups, and I had never collaborated like that before: arguing my way through problems with three or four or five other people into the wee hours of the night. We were a small band of students giddily, exhaustedly trekking through an abstract moonscape, helping one another across patches of ice or fighting over which direction to head next. The egos, the insecurities, the unabashed nerdiness! I miss it still.

Also, at nineteen, so much is up in the air, open to question, unreliable. I think part of what I liked about math was simply that it seemed like a sure thing, as sure as a thing could be, a solid mass of true and rigorous and irreproachable knowledge that I could grab like a pole on a bus.

See, it is 56. You will find it right.

I'll just go ahead and say, in case it's not already clear, that André can be pretty abrasive. He is arrogant, impatient, short with people. Wound tight.

As the house has no electricity and hence no electric fans, a boy is paid a pittance to stand on the veranda and pull at the string that sets in motion the *panka*, a piece of cloth hung from the ceiling, so that the air might circulate while André naps. Sometimes the boy dozes off, and André wakes up in a sweat and shouts "*Pankevale!*" to rouse him.

(His sister not only would've refused to nap under those conditions but surely would've insisted upon yanking the string herself, while someone else—the boy—slept.)

A caustic sense of humor. A hatred of flattery.

But he loves India. Everything is brighter or else darker than it is in France, louder or else more opulently silent, more fragrant or more foul. Oleander bursting in front of the house, mangoes rotting in the back. He loves it. He loves the spicy food. He loves to read the railway timetable in bed and take trips on the weekends.

On a full-moon night he and two friends are lent a car and driver by a university benefactor, and they travel to Fatehpur Sikri, where in the latter part of the sixteenth century the

Mughal emperor Akbar built magnificent palaces of red sandstone for himself and his courtiers, only to abandon them because of a lack of water. There are no gates or guards or hours of operation, and under the light of the moon André and his friends wander through the abandoned imperial city, through courtyards and galleries and harems where, in the blazing moonlight, lattices within the windows cast honeycombs of shadow on the floors.

Another time, he goes by train to a nearby village, and when he returns to Aligarh he finds the station strangely deserted, without a single employee in sight. Finally, in what passes there for a toilet, he finds one of them hunched miserably over the latrine. Later he'll discover that the chemistry professor, determined to prove his hypothesis in the case of the missing guavas, traveled to the orchard and injected the entire shipment with a strong purgative. In this way the professor identifies the thieves as the railway employees he suspected all along.

Then there was the fact that I had a serious boyfriend for the first time, I was really in love for the first time, and my elation seeped into everything, I saw the world through infatuation-tinted lenses. Part of loving math, for me, was loving a person who also loved math, who walked with such long strides, at once forceful and awkward, and called bullshit whenever he saw it—which was often—and in whose company I felt let in on a truer and more powerful and more beautiful mode of being in the world, even if in retrospect it seems to me that sheltered as we were inside the concentric bubbles of our relationship and math and college itself, we were barely in the world at all.

I mimicked the ironic but heartfelt praise that he and his roommates would give some new revelation from a math or science class: *Dude*, that is so *rad*.

The earliest recorded Hindu math makes its appearance in the Sulvasutras, which give rules for the building of altars. A design may incorporate squares, circles, and semicircles, but every shape used in an altar is required to have the same area. Thus every altar is a geometry problem.

Ancient Greek legend gives us another altar exercise. According to Eratosthenes and Theon of Smyrna, who both recorded the story, the city of Delos was once afflicted by a plague, and its leaders traveled to the oracle at Delphi to ask what they might do to halt the epidemic. The oracle—supposedly speaking for Apollo but sounding more like a math teacher than a god—instructed them to double the volume of Apollo's altar, which was in the shape of a cube. Hence the problem of doubling a cube.

"Astonishing Phenomenon," André begins a letter to his sister, in November 1931. "Word of your exploits has reached me here."

Simone has by now graduated and become a lycée teacher in the town of Le Puy, southwest of Lyon. She's also made herself an advocate for the city's jobless men, whom she has observed from the windows of the girls' high school, as they break stones in the Place Michelet. Grubby and gaunt, swing-ing scratched-up hammers—all day she can hear the thwacks and clanks, and every so often the cry of someone who's been hit in the foot or has thrown out his back. If the men work the entire day and produce enough, they earn six francs from the town. Simone accompanies them to meetings of the city council and mayor, to ask for better work and pay.

She is described in a newspaper report as a "bespectacled intellectual lady, with her legs sheathed in sheer silk," not that she's ever worn fancy stockings or fancy anything. The "miserable, unemployed worker" deserves our sympathy, notes the article's author. "It is for him that such feelings should be reserved and not for those intellectuals who want to 'make a splash' and who flourish on the misery of the poor like mushrooms on humus."

"Mushroom on the Humus," begins another letter from André. "I send you my wholehearted congratulations and encourage you to continue down this road. *Unqilab Zindabad!*"—Urdu for *Vive la révolution!*

The Hindus were the first culture to use negative numbers. It's sometimes said that they discovered them, though it's un-likely that they thought of them as a discovery, that is to say as something new they'd unearthed about the essence of number (if number can be said to have an essence). Evidently they considered negatives more of a trick, an accounting de-vice: when negative numbers first appear, in the work of Brahmagupta, circa A.D. 628, they represent debts, as against positive-valued assets. It took centuries for them to be seen as legitimate numbers. Even five hundred years later, by which time, you might suppose, negatives should've settled unremarkably into the mix, the Hindu mathematician Bhaskara notes that although one may find a negative solution to a problem, "people do not approve of negative solutions."

When recording an equation with more than one unknown in it, the Hindus used the same words they used to denote colors. Writes Morris Kline, a historian of mathematics: "The first one was called the unknown and the remaining ones black, blue, yellow, and so forth."

The unknown plus the square of blue minus three times red equals zero.

Kline again: "It is noteworthy that they found pleasure in many mathematical problems and stated them in fanciful or verse form, or in some historical context, to please and attract people."

"Dear Noumenon," Simone writes back. "Thank you for the congratulations and encouragement . . . It's been established here that I am an agent of Moscow."

Agent of Moscow, because another article has called her the "Red virgin of the tribe of Levi, bearer of the Muscovite gospels," though in fact she's skeptical of the Communist Party (as she is of all political parties) and has soured on Stalin, well ahead of many of her left-leaning friends. But since she writes articles for radical newspapers, she is thought to be a Communist and is followed to school by police.

Noumenon, or a thing whose existence can be reasoned but never perceived—like God, like the soul. Like mathematics? As opposed to phenomenon, an appearance, a thing apprehended by the senses.

The sister who could be perceived directly. The brother a distant god.

Un qil ab Zindibad!!! she signs off.

I remember walking home from the Science Center after midnight, a layer of new snow underfoot. No wind, no one else around. I slipped into an enfolding stillness. Although it was the darkest part of a winter night, the streetlamps were ablaze and the snow was shining; it didn't seem dark at all but like I was walking through a lit passage back to my room, a tunnel of light cushioned by an endless black sky. I experienced then, experienced from time to time, a kind of pleasure that came only after having thought hard about math, the mental equivalent of having gone for a long run. A gentle euphoria.

Negative numbers infiltrated Europe during the Middle Ages, were imported like viruses by international voyagers, who brought from the Middle East certain Arab mathematical texts that elaborated on the advances made by the Hindus. Yet in the sixteenth century, people still questioned whether negative numbers were truly numbers. Michael Stifel called them "absurd numbers." Blaise Pascal said they were utter nonsense. John Wallis reasoned that they must be larger than infinity. And Gerolamo Cardano—a.k.a. Jerome Cardan, a notorious scoundrel, professional gambler, physician, and caster of horoscopes, who happened also to be an exceptional mathematician—declared them impossible solutions. "Fictitious numbers," he called them.

Negatives weren't the only shady characters. There was a whole rogues' gallery of irrationals, like $\sqrt{2}$—they kept poking their heads up in equations, but were they numbers? "We find that they flee away perpetually," Stifel wrote, before concluding that an irrational was "not a true number, but lies hidden in a cloud of infinity."

Worse yet, along came the square roots of negatives—Descartes gave them the name "imaginary numbers"—and the hybrid creatures we now call complex numbers, composed of a real number added to an imaginary number. Cardano associated them with "mental tortures." A complex root of an equation is "as refined as it is useless," he wrote.

André's mathematical investigations dead-end, and dead-end again: He tries to expand upon the work he did for his thesis, on Diophantine equations, without making any progress. He has an idea about making use of John von Neumann's work on unitary operators in Hilbert spaces to attack the problem known as the ergodic hypothesis, but the idea isn't specific enough. He flirts with celestial mechanics, drops it.

In the spring he orders the boy servant to drag his bed up to the roof terrace. It's almost too bright to fall asleep there, the tropical sky is so clear, the stars sublime, but truth be told he has had just as much trouble falling asleep inside the house. He reads the railway timetables restlessly, he's a young man with no wife, no girlfriend, in a foreign country, and so naturally he has a lot of pent-up energy that hurls itself into his weekend sojourns.

Might there have been some girl in Kashmir? A Dutch lady traveler on one of those trains? Some colleague's cousin, who mocks him but later slips him a note? Or was he untouched, untouchable for those two years? He lies up there on the roof and moans at the stars.

The madness of reason.

On other winter nights I would work in my dorm room, and when it grew too cold I would run my electric kettle as a heater, boiling the water down and refilling it and boiling more water until I'd made a sauna of the area around my desk and futon, until the lone window wept. I would sit there, in a fog of my own making, trying to demonstrate small truths. Prove this, disprove that, describe such and such explicitly.

In her keenness to inhabit the life of a worker, Simone wangles a visit to a coal mine, normally off-limits to women. There she is not only escorted down into the shaft but allowed to try her hand at the compressed-air drill, a deafening machine that sends continual tremors through her small body. She holds on for dear life. Had someone not stopped her, a companion later reports, she would've kept on using the air drill until she collapsed.

She asks the boss to hire her.

He declines.

The coal miner, she writes afterward, is a pawn in a titanic struggle between coal and compressed air: "Clinging to the pickax or drill, his entire body being shaken, like the machine, by the rapid vibrations of compressed air, he confines himself to keeping the machine applied at each instant to the wall of coal, in the required position." The miner winds up becoming a part of the machine, "like a supplementary gear."

She asks, under what conditions could a revolution possibly be successful, given that the machinery of labor is itself oppressive? It would have to be a technological revolution, as well as an economic and political one.

What André most desires, during those long nights in India, is to set his own head spinning. He once, during a stay in Germany, entered a mathematical fugue state that lasted for hours and lit the way for his dissertation, but he doesn't know whether it will ever happen again. "Every mathematician worthy of the name," he'll write in his memoir, "has experienced, if only rarely, the state of lucid exaltation in which one thought succeeds another as if miraculously, and in which the unconscious (however one interprets this word) seems to play a role."

At last comes the spark, a new idea about certain functions. Functions of several complex variables—in the realm of Cardano's mental tortures. It hits him on his rooftop bed as he is half asleep, a silk-threaded structure spinning its way out from a worm in his mind, and shocks him awake, he rolls onto his stomach and gropes for the timetable, wanting the stub of a pencil he tucked between its pages. The pencil rolls away, and now he is down on his knees, patting the tiles ever so gently in hopes of finding the pencil and not a scorpion. At last there it is, and he starts to write over the schedule of the Delhi–Calcutta line. His ideas arrive as clear and bright as the sky, there are routes among the stars he never noticed before, it's as though he doesn't have to produce the thoughts any longer, they are simply supplied to him, and he is the scribe who turns them into symbols. His mind has been razed and there are barely any words in it. Only gratitude, only functions.

Does the value lie in what he's writing down or in this moment of lucid exaltation itself, this obscure bliss? Could it be a new porthole on reality, the theorem that is taking form, or is it better considered as a kind of access to the innermost architecture of thought? An infinitely nested diagram, unfolding itself. Arguments and computations, functions suspended between floating fields of numbers. Cool flames. The mathematician's early harvests.

The legend of Archimedes, struck by insight during a bath and then running naked through the streets of Syracuse, shouting, *Eureka! Eureka!*

"I have reviewed the circles on the basis of the demonstrations you have given me," writes a young worker to Simone. When she's not teaching or going to union meetings, she volunteers to give free lessons, convinced as she is that work-ers, in order to advance, must be better educated—and that geometry lies at the heart of a proper education.

"In the three lessons I had with you, you have given me almost all the elementary facts of geometry; it is a pity that I cannot see you more often, for I would have ended by be-coming a truly learned person," the letter continues. "With you as the teacher I was never bored for a second; and these few instants exalt all the noble thoughts that inhabit me. If I could see you more often, I would make double progress, intellectual as well as moral."

I wonder whether this young man was in love with her, can't help wishing that he was and that she would've loved him back. When Simone was a child, her mother had a strong fear of germs and obsessed over hygiene. Simone absorbed that worry and exaggerated it to the point that she shrank from others' touch. She didn't hug or kiss people. As far as anyone knows, she never had a lover.

So there's no telling whom she might've been thinking of, if she was thinking of anyone, in this passage from her note-books: "All our desires are contradictory, like the desire for food. I want the person I love to love me. If he is, however, totally devoted to me he does not exist any longer and I cease to love him."

Cardano, by the way, was one of those men of the Renais-sance whose polymathic, credulity-straining lives gave us the whole notion of the Renaissance man. He was an outlandishly brilliant thinker and, it seems, a total dick: "high-tempered, devoted to erotic pleasures, vindictive, quarrelsome, conceited, humorless, incapable of compunction, and purposely cruel in speech," writes Kline in his history. Born in 1501, the illegitimate son of a lawyer and a woman who'd tried and failed to abort him, Cardano was imprisoned for heresy because he'd committed the ecclesiastical faux pas of casting a horoscope for Jesus. Yet after he finished his prison term, the pope hired him as an astrologer.

In addition to gambling, playing chess for money, and practicing medicine, Cardano would, for a fee, detect your character and fate based on your facial irregularities. He wrote an entire book on this kind of conjecturing, with some eight hundred labeled diagrams of faces.

"A great enquirer of truth, but too greedy a receiver of it," Sir Thomas Browne would later say of him. During his lifetime Cardano published 131 works, encompassing not only mathematics, astronomy, physics, and medicine but also astrology, dreams, portents, and charms. He wrote on angels and demons. He plagiarized his father's friend Leonardo da Vinci. He wrote a memoir, *De Vita Propria Liber*, in which he lamented his wretched boyhood and the extreme poverty he endured as a young man. His son, Giambatista, was executed for poisoning his wife.

Cardano died on September 20, 1576, the very day he'd predicted for himself. In his final year, he had fourteen good teeth.

The more than seven thousand pages he left behind are a monument to the encyclopedic tendency, to the idea that a powerful intellect could be in possession of the whole scope of human learning. When did that idea expire? Now there is too much to know, not to mention the whole problem of the unreliability of the knower. Now we have machines for knowing.

Or we turn knowledge into an ornament, a bauble. There's a welcome tingle brought on by the flotsam of scholarship, the odd facts that wash up onto the shore. The fourteen good teeth.

Simone works in defiance of her own body. She goes for long stretches without food and rest, and that's not the only way in which she flirts with extinction. She teaches a course for union members called Insights into Marxism, and her students fear for her. "This Simone," complains one man attending the class, "just look, five times she lights her cigarette and throws the matches and sparks on her blouse. She'll end up by setting herself on fire."

André returns to Europe in May 1932 and stops in Rome to pay a visit to Vito Volterra, a distinguished older mathematician. After he explains to Volterra his progress in functions of complex variables, the Italian abruptly stands and runs toward the back of the apartment, calling to his wife, *Virginia! Virginia! Il signor Weil ha dimostrato un gran bel teorema!*

Virginia, Virginia, Mr. Weil has demonstrated a very beautiful theorem!

4. THE WORD *CONJECTURE* DERIVES FROM a root notion of throwing or casting things together, and over the centuries it has referred to prophecies as well as to reasoned judgments, tentative conclusions, whole-cloth inventions, and wild guesses. "Since I have mingled celestial physics with astronomy in this work, no one should be surprised at a certain amount of conjecture," wrote Johannes Kepler in his *Astronomia Nova* of 1609. "This is the nature of physics, of medicine, and of all the sciences which make use of other axioms besides the most certain evidence of the eyes." Here conjecture allows him to press past the visible, to sacrifice the certainty of witnessing for the depth and predictive power of theory. There's another old definition of *conjecture* that means something inferred from signs or omens (for example, from a Renaissance work on occult philosophy: "Whence did Melampus, the Augur, conjecture at the slaughter of the Greeks by the flight of little birds . . .").

Elsewhere it's hokum, claptrap, bull: "Conjecture, which is only a feeble supposition, counterfeits faith; as a flatterer counterfeits a friend, and the wolf the dog," wrote one early Christian theologian. So it's a word with contradictory meanings, since at times conjecture carries the weight of reasoning behind it, and at other times it's a wild statement, an unfounded claim. Good thinking or bad, clever speculation or a reckless mental leap.

In contemporary mathematics, conjectures present blueprints for theorems, ideas that have taken on weight but haven't been proved. Couched in the conditional, they establish a provisional communication between what can be firmly established and what might turn out to be the case. More than a guess, conjecture in this sense is a reasoned wager about what's true.

A rough draft. A trial balloon. It seems to me laced with optimism, a bullishness about what could, in the future, come more fully to light.

André, now a professor at the University of Strasbourg, travels back to Paris and convenes a meeting of young mathematicians at the Café A. Capoulade. They all teach similar undergraduate courses in analysis—higher-level calculus—and they are dissatisfied with the existing textbooks, so they've decided to form what they call the Committee for Writing a Treatise on Analysis.

This rather dry mission soon takes a strange turn. The Committee for Writing a Treatise on Analysis will evolve into a semisecret society, a sort of wry mathematical mystery cult.

My math fever dream lasted for two and a half years, which were spent messing around on the lower rungs of a tall ladder that stretched into the clouds, that led to a cloud land of tantalizing abstract structures, curves and surfaces and fields and vector spaces, accessible only to those who learn the elaborate cloud language, a vehicle for truths that cannot be expressed in any other tongue.

Then I stepped off the ladder and walked away. It's not often that I experience even a passing wish to go back, and even in those moments the allure of math is much fainter than it was in college, since now I have no illusions that I would ever make it very far up—I'm left to imagine that land, and what I wish for now is less the specific math knowledge than a certain constellation of feelings that came with it.

Simone goes to work in a factory because she wants to investigate directly what she hasn't been able to figure out theoretically, namely: How can an industrial society be organized in a way that its workers are not oppressed? And then there's that self-mortifying impulse, the fact that she has wanted to do hard physical labor since she was a teenager. She tells her friend Simone Pétrement (who will later become her biographer) that although she's scared—she is notoriously clumsy, not good with her hands—she is determined to kill herself if she can't manage the work.

She is twenty-five. Her confidence, Pétrement would write, "was quite terrifying, especially when one knew her almost inhuman energy and her lack of self-pity."

Into a den of machines. A woman regularly overcome by headaches assigns herself to the factory floor with its stamping press, the fly press, the iron crank, the compressed-air hose, the screws, the blades, the mallets. The screech and whine and hammer and hiss. She is assigned to fit copper plates into magnetic circuits and must take care not to ruin the pieces in the process.

The days are short, and when the windows go dark the wan lamplight hardly compensates. She can barely see. It's loud and it's dark and she can't keep up with what the foremen demand, the required rate of work. Holding a flashlight in one hand and adjusting newly assembled parts with the other. "Manual labor," she'll write later. "Time entering into the body."

Wednesday, she doesn't make the rate. Thursday, she doesn't make the rate.

She imagines that the others pity her, but then she sees how they side with the foreman when another woman is fired. The fired woman was tubercular, her husband unemployed—but she botched a job and hundreds of pieces had to be done over. She ought to have known better, the others comment. "You've got to be more conscientious when you have to make a living," they say. Even on the night shift, even in the near darkness.

"What I went through there marked me in so lasting a manner that still today when any human being . . . speaks to me without brutality, I cannot help having the impression that there must be a mistake," Simone will recall.

She is assigned to a small workshop, separate from the rest of the factory, and instructed to insert copper bobbins into a furnace and then take them back out again. She burns her hands and her arms, comes out blistered and scarred. Yet it's her favorite part of the factory, because the workers there are decent to one another.

The young mathematicians hold their first conference in the summer of 1935, near Lac Pavin, a volcanic crater collared by pine forest in central France. Seven months after forming the Committee for Writing a Treatise on Analysis, they've advanced well beyond their original intention to improve undergraduate math education. This has been supplanted by a much grander goal, no less than laying a formal, consistent, and comprehensive foundation for all of modern mathematics—or at least all of it that they find interesting.

Taking a break from a stalled discussion of analytic functions, several members of the group flee to Lac Pavin and dive naked into the water, yelling out a Greek name: *Bourbaki! Bourbaki! Bourbaki!* Over and over again, their shouts echoing across the surface of the lake: *Bourbaki! Bourbaki! Bourbaki! Bourbaki!*

One seed of this happy outburst was planted in 1923, back when André and many of the others at the conference were studying at the École Normale. A third-year student named Raoul Husson played a prank on the first-years by posing as one Professor Holmgren. Wearing a false beard and speaking in an indefinite foreign accent, he gave a lecture on a series of made-up results, which culminated in the presentation of "Bourbaki's theorem."

And, when he was living in India, André had advised a

young mathematician friend by the name of Kosambi, who was caught up in an academic rivalry with another man, to flummox his competitor by publishing an article about an imaginary Russian mathematician with a Greek name. Kosambi did just that; his (spoof) paper "On a Generalization of the Second Theorem of Bourbaki" appeared in the *Bulletin of the Academy of Sciences of the United Provinces of Agra and Oudh.*

The group at the summer conference adopts the name Bourbaki, a mischievous pseudonym that also serves a purpose, in that they'll avoid having to sign every single one of their names to publications. In the fall, André will submit to the journal *Comptes Rendus* a perfectly serious article concerning the theory of integration, yet he'll report that it was passed on to him by a man named Bourbaki. "I'm sure you'll recall that Mr. Bourbaki is the former professor of the Royal University of Besse-en-Poldévie whom I met some time ago at a café where he spends most of his day and even the night, having lost both his job and most of his fortune amid the troubles that caused the unfortunate Poldévian nation to disappear from Europe," André will write to the journal editor, a fellow mathematician who is in on the joke. "Now he earns his living at the café by giving lessons in belote, the card game he plays so brilliantly."

There was, by the way, an actual Charles Bourbaki, a nineteenth-century French general of Greek heritage who distinguished himself in battle and as an inspector general of the infantry before he was sent, in 1870, to command a miserable, half-starved French army already losing to the Prussians. After a defeat at Héricourt, in eastern France, he was forced to retreat through Switzerland, where his soldiers' guns were confiscated and he tried to kill himself. The attempt failed, and he eventually returned to France.

The brief Kafka story "The Top" is about a philosopher who loiters around children because he is fascinated by their tops and hopes to catch one in mid-spin, convinced that a spinning top could lead him to enlightenment—that "the understanding of any detail, that of a spinning top for instance, was sufficient for the understanding of all things."

As a top spins, the philosopher goes "running breathlessly after it," full of hope, but upon catching it he is disappointed, for "when he held the silly piece of wood in his hand he felt nauseated." It's the pursuit, the running after knowledge, that takes his breath away.

Commenting on this story in the preface to her book *Eros the Bittersweet*, Anne Carson doubts that the desire to gain knowledge is what really motivates Kafka's philosopher. "Rather," she writes, "he has become a philosopher (that is, one whose profession is to delight in understanding) in order to furnish himself with pretexts for running after tops."

Running *breathlessly* after tops, in that heightened state that comes of being deprived not only of the thing you're chasing but of the air you need to chase it. Sometimes I think of writing in this light, too: as running after tops.

A partial list of women Simone encounters at the factory, as recorded in her notebook: Mme Forestier, Mimi, Cat, Mimi's sister, a coworker on the iron bars (Louisette), a blonde from the munitions factory, a redhead (Josephine), a divorced woman, a mother of a burned child, a woman who gave her a roll to eat, an Italian woman.

Some of the men: a violinist, a conceited blond, an old man with glasses, the singer at the furnace, the worker in drilling goggles, the boy with the mallet, a young blond Italian, a welder, a coppersmith.

The grooves in the forehead of her supervisor, Mouquet, look to have been carved there by one of the machines. He is tormented by invisible forces and torments the workers in turn. One day Simone is told to redo a long metal-polishing job, which makes her furious. Is the demand justified, or is it bullying? She can't even tell. The way Mouquet instructs her to go about it, she has to duck a counterweight that swings toward her every time a piece is polished.

Hindering her productivity, she believes, is her contemplative nature. "I am still unable to achieve the required speeds, for many reasons: my unfamiliarity with the work, my inborn awkwardness, which is considerable, a certain natural slowness of movement, headaches, and a peculiar inveterate habit of thinking, which I can't shake off."

She feels disgusted by the exhaustion and by how little she'll earn from her slow work. She's disgusted by herself, her identity worn down as though in the polishing machine. Oppression, she concludes, doesn't foster a spirit of rebellion but a kind of docile slavery.

And one's inner life, she writes in her notebook, is merely a kind of temptation. One shouldn't indulge in emotions unless they are useful. Can a person experience feelings without imagination, without projecting them into the past or the future? This seems to be her aim. "Above all, never allow oneself to dream of friendship. Learn to reject friendship, or rather the dream of friendship."

Determined to live off what she earns, she insists on paying her parents when she dines with them. Her mother responds by stashing coins around Simone's apartment, knowing that Simone is absentminded and won't realize she didn't leave them there herself.

"She's killing herself, and she doesn't listen to me," her father complains.

"It is not by chance you have never been loved," she writes in her notebook.

André's memoir doesn't offer any details about how he met and wooed his eventual wife, Eveline, perhaps because when the two of them became lovers she was married to—and had a son by—one of his colleagues, René de Possel, another of the original members of the Bourbaki group. The story of how André courted her is, must be, a story of betrayal.

Let's say it happens at that first conference, in the summer of 1935. Young professors wandering through fields of wildflowers, sunning themselves in chairs. Mountains in the distance. He and she cross paths in the dining room, in the garden. She is short and broad-shouldered, curvy, with wide eyes and an easy smile. Maybe she always wanted to marry a genius and has not quite found one in her first husband. Is it that, André's intellect? Or maybe his genius in combination with his nearsightedness, the way he is so cocky but can't figure out where he's left his pen. Wiry in a way that makes him look taller than he is, worldly, imperious and mocking around the other men, but when she finally works up the nerve to say hello one morning, he blushes. He stares at her as he talks to her.

Stray hairs blowing forward, into her mouth, distract him.

He isn't wearing a tie, and she likes his open shirt, his skin.

Words come and go, which is to say they speak them while taking the measure of each other's bodies, shyly looking and then looking away.

He isn't wearing socks either, she notices.

He eyes one of the silver buttons of her blouse, thinks about those buttons all afternoon.

I suppose I'll see you at dinner, he says.

It was good to meet you, she says.

As he walks away, he stumbles over his own feet.

A second-rate mind like de Possel doesn't deserve her—that's what he decides. Over the next few days he keeps a furtive eye on Eveline and René and convinces himself that there is no real warmth between them, only schedules and habits, dry kisses on the cheek.

Within days André has lured Eveline out to the lake. Sneaking off in the dark, they ride bicycles to the edge of the crater, where the lake is ritzy with starlight, and he tells her this and he tells her that, how much brighter the stars are in India, let's say, and then next thing they know they've entered the water. Her camisole becomes transparent, an aquatic creature daubed across her heavy breasts. Their wet skin, the warm wet rippling night. Afterward, on the shore, she wants to wait until her hair dries before they go back. As though wet hair were the only giveaway.

In eight months of line work at two different enterprises, Simone finds no answer to the problem of social organization—that is to say, she concludes that there is no answer, that she has become a slave among slaves. By the time she leaves the factory for good, she's utterly spent. She hatches a plan to travel by freighter down the Spanish coast, but along the way her health declines, and her parents take her to a seaside town in Portugal to rest. There, one evening as she wanders through the village alone, she has the first of several mystical encounters that impel her toward Christianity.

For the festival of the local patron saint, the village women have formed a procession. They wend their way around the docks, carrying candles and singing hymns permeated with "heart-rending sadness," as Simone will later write. A thought—a conviction—comes to her: that Christianity is the religion of slaves.

She will say that not only the Christian religion but elements of other religious traditions and the beauty of the world and its reflection in great works of art delivered her "into Christ's hands as his captive." She becomes an idiosyncratic almost-Catholic, at once mystical and scholarly, skeptical of the church and declining to be baptized but believing herself possessed by Christ, believing the invisible world to be more real than the visible one.

————

Carson's *Eros the Bittersweet* takes up the idea that when ancient Greece became literate, developed a culture of writing and reading, which set in around the seventh or sixth century B.C., its collective mind was blown. Literacy comes with its own psychology, Carson suggests: to become a reader you must learn a new kind of self-control. You have to shut out the world in order to focus your attention, and in so doing you develop a new awareness of the interior self.

Distant objects or people, represented by symbols on a page: we are so accustomed to this that it's hard to conceive of a time in human history when it was a new phenomenon. Carson compares this representing, the conjuring of things by written words, to the way that a lover constructs a mental image of an absent beloved. Desire spans the distance between the abstract thought and the actual person. Everything is triangulated—the lover, the beloved, the image. The writer, the thing, the word.

"A mood of knowledge is emitted by the spark that leaps in the lover's soul," she writes. "He feels on the verge of grasping something not grasped before." It's not the knowledge itself, not consummation but the mood, the excitement when you are *on the verge of grasping.*

Skeptical though she is of political idealism, Simone never relinquishes her passion for self-sacrifice. In 1936 she travels to Spain to join the republican fighters in the civil war, less out of devotion to the cause than in the belief that once war cannot be prevented, one must submit to one's share of misfortune. Posing as a journalist, she reaches the front and manages to join a small international commando group.

Simone is nearsighted and inept with a rifle, terrifying the others every time she has the weapon in hand. Her captain assigns her to kitchen duty, but she still finds herself, one afternoon, stretched out on her back while nationalist planes fly overhead, gun pointed toward bombers in the sky. Once again she's put herself in a situation that's nearly intolerable yet brings its own hard thrills. Absolute visibility: she's pierced by the sight of the summer sky as she lies there, grappling with a heavy gun she can barely operate and beset by a fierce headache, by beauty and terror and pain. The planes are too high to shoot, and the soldiers scramble back to camp.

Even the kitchen has its hazards. The comrades start their cooking fire in a hole, so that the flames won't give their position away, and one morning Simone steps right into the hole, her leg landing in a pot of boiling oil. As another woman removes Simone's stocking, her skin peels off with the wool. She is taken to Barcelona and then to a hospital south of the city, which likely saves her life, as several weeks later most of the international group, including all of its remaining women, are killed in an engagement at Perdiguera.

During the siege of Syracuse by the Romans, circa 212 B.C., a Roman foot soldier encountered Archimedes at the shore, studying a diagram he'd drawn—according to one account, he was so engrossed by a problem he hadn't even registered the Roman invasion. The soldier ordered Archimedes to follow him, but he was seventy-five by then, and maybe he didn't hear so well, or maybe he told the soldier that he would come along once he finished working through the problem. Although the Roman command had put out word that the famous mathematician should be spared, the angry soldier slew him on the spot.

Strategies for tackling problems, from Pólya's *How to Solve It*: Do you know a related problem? Look at the unknown! Here is a problem related to yours and solved before. Could you use it?

But what about the problem of too many related problems? My weakness for juxtaposition: I'll sense that one thing might be illuminated by another thing and go chasing the other thing (Thing Two, like one of Dr. Seuss's little devils in *The Cat in the Hat*) off in another direction. For better or worse, a light paranoia goads me along. Maybe it's all connected! This and this and this and—look, over there—that. There's the bringing together of disparate elements that informs a conjecture, and then there's the mental nausea brought on by the fact that there's too much out there to know. Not grasping but googling. I can't always tell one from the other.

The ancient world left a record of at least one female mathematician: Hypatia, the daughter of Theon of Alexandria. She wrote commentaries on the mathematical works of Diophantus and Apollonius and then died in A.D. 415 at the hands of Christian fanatics who, because she refused to give up her pagan beliefs, dragged her through the streets and scraped the flesh from her bones with pottery shards or possibly—the relevant Greek word has more than one translation—with oyster shells.

Simone attends Bourbaki conferences, though she can't follow most of the technical discussion. A photo taken in 1938, in Dieulefit, shows her standing in front of a doorway with her brother and five other mathematicians. She is the only one looking away from the camera, as though she knows better than to claim a place there.

The Bourbakians often yell back and forth, cut one another down. Shut up! No, you shut up! During one of their meetings, at a hotel in the French Alps, an employee wonders whether to call the police after hearing one of them threaten to throw another out the window.

She does her best to keep up, smoking, thinking, pacing back and forth in the rear of the room while the men attack their math, attack one another. During the breaks she hits up André with questions, while he stirs sugar into his coffee, maybe wishing a little that she would go home and at the same time finding a familiar satisfaction in continuing to teach his first pupil. (Besides, what home does she have?) Then again sometimes her questions jab serendipitously in the right direction, cause him to reconsider some ignored assumption. To the others it's startling to see his same glasses, his same face attached to this body clothed in an unstylish dress and an off-kilter brown beret, carrying on in that odd monotone as she argues, via the château's telephone, with the editors who publish her political articles.

Although André's work is beyond her, Simone still thinks that a better social order depends somehow on making mathematics accessible to the masses. "Mathematics above all," she writes. "Indeed, unless one has exercised one's mind seriously at the gymnastic of mathematics one is incapable of precise thought, which amounts to saying that one is good for nothing."

Good for nothing! And yet her life as she lives it during the 1930s, trying and failing to find a toehold in practical activity, suggests a contrary principle—that too much knowledge, too much bowing over Descartes, leaves a person incapable in other respects.

Sometime late in the decade, her father reads an article in a medical journal describing the symptoms of larval sinusitis, which remind him of symptoms (the frequent headaches, the congestion) that Simone has. He wonders whether she could have somehow contracted this condition, even though larval sinusitis is primarily a disease of sheep, in which the larvae of a particular kind of fly make their way into an animal's nostrils. The reported human cases have occurred mainly in shepherds.

Nevertheless, Simone reads the article herself, agrees it might be possible, and tries the recommended treatment: doses of cocaine applied to the sinuses.

For all their lofty intelligence, the men behind Bourbaki are goofballs: They adamantly maintain the fiction that their papers and books are the work of a man named Bourbaki, member of the Royal Academy of Poldévia. They refute reports that the name Bourbaki refers to a group in France. After Ralph P. Boas, the editor of the American journal *Mathematical Reviews*, writes a paragraph about Bourbaki for the *Britannica Book of the Year*, explaining that N. Bourbaki is not a man's name but a pseudonym, the group propagates a rumor that it is Ralph P. Boas who does not exist—that B.O.A.S. is in fact an acronym for the cohort that edits *Mathematical Reviews*.

Playfulness aside, their project is to systematically ground all of contemporary mathematics in a series of abstract, rigorous volumes titled *Éléments de mathématique*. This endeavor is a distant cousin to one undertaken earlier in the century by the British logicians Bertrand Russell and Alfred North Whitehead, who tried to put all of mathematics in the form of a consistent logical system—only to see Kurt Gödel, a brilliant Austrian, pull the rug out from under them by showing that any such system would contain the propositions that couldn't be proved within the system. But Bourbaki doesn't want to build the same kind of logically watertight apparatus. Its members, working mathematicians, don't tend to care that much about the logicians and their fancy tricks, those snakes swallowing their own tails. Bourbaki aims to put everything on the same footing, to unify it by way of an axiomatic language. To become the Euclid of the twentieth century.

They meet and hash out the contents of a volume, assign to one or two members the task of writing it, then meet again and subject the draft to the harshest criticism. As textbooks they aren't terribly useful. More than a hundred pages of set theory must be laid out before Bourbaki can define the number 1. But they conjure a vision of modern mathematics, a vision of structures and maps, according to which the intrinsic qualities of mathematical objects (numbers, sets, spheres, or what have you) matter less than the relationships among them.

They find a congenial publisher in Enrique Freymann, a descendant of German immigrants to northern Mexico, who was a painter and a diplomat before he married the daughter of a French publisher and agreed to run the company out of a cluttered office near the rue de la Sorbonne.

A squat man, let's say, who loves to talk and loves to read and is fascinated by science. He drives a big Renault convertible, badly, around the streets of Paris. He is as surprised as anybody when the company, under his direction, begins to turn a profit. He spends the better part of his days perusing the rare books he picks up during his lunch hour from the stalls along the Seine; or he rambles on to whoever walks in, addressing miscellaneous themes, many of which eventually lead him back to his desolate yet eventful childhood in a Mexican town near the U.S. border during the volatile years after the revolution. Or else he tells of his initial puzzlement by and subsequent education in French decorum and social hierarchies, or he'll speak of the irony of having traded the dust of the Chihuahuan desert for the dust of an out-of-the-way publishing office.

But good God, did Simone Weil really have tiny worms lodged in her nose? Hard to take that seriously, it's like a story one kid tells another kid on the playground. Yet it seems she and her father thought that it was possible—and that the remedy was cocaine.

Her attraction to suffering was deep, and some of her experiences beggar belief. I'll be reading along in a biography of her, and everything is what it is, but when I close the book and try to imagine what I've just read, or later on, while I'm driving maybe, I'll remember what I learned earlier and be baffled by it. Wait, *what*?

Toward the end of his life André would be called "the last universal mathematician" by *Scientific American*, which is to say the last to work in so many of its subdisciplines, though by then it may have been too late for anyone to really be universal in scope. You could, at least, call him a descendant of the universalists, bearer of the deteriorating mantle of David Hilbert and Henri Poincaré—turn-of-the-century omnivores who are also referred to as the last universalists. Like them, André took a broad view, probing for connections linking different parts of mathematics.

Or you could instead make a case that the last universal mathematician, the last to really have an aerial view of the field, was not any one person but the group known as Bourbaki. With them the dream of the encyclopedia survived: Knowledge as a neat set, a series of ordered volumes. Fine print and thin paper. A committee of surveyors taking the measure of the kingdom.

———

It's not as though André lives in a cave of pure abstraction. He buys the newspapers and follows the mounting threat of war (how could he not?) and scorns the leaders of France, whom he'll compare to sheep, "resigned to follow wherever they were led—be it to the slaughterhouse." Yet there are days when he dives into his mathematical work, loses himself in it, and maybe it's that tendency that inspires Simone, one day in 1939, to send him a telegram that says, RECOMMEND READ NEWSPAPERS. The occasion is Hitler's invasion of Czechoslovakia. The war has begun.

PART TWO

5. HOW I WOULD LIKE TO WRITE SOMEthing as clean and powerful as the best kind of mathematical proof. In pen, on quadrille paper: lines of black script conforming neatly, inevitably, to the faint blue squares. The sun would sly its way out from behind the clouds and beam right at your head, warm up some inner lobe of the walnut. A hint, a glimpse into the nature of things.

André Weil is arrested in Finland, on suspicion of spying, in December 1939. He's been in Scandinavia for months, initially on vacation with Eveline—a trip they take to escape the impending storm at home—and then prolonging his stay after France enters a war in which he doesn't intend to fight. The army is not his destiny, his dharma, he's convinced, and so he feels justified in avoiding the draft. He has taken from the Bhagavad Gita an idea that everyone has a purpose in life, and his is mathematics, he believes, not soldiering.

He doesn't realize, at this stage, that he is entering into a life of professional exile—but then again, would he behave any differently if he knew?

When the Finnish police take him into custody, he expects that he'll be questioned and released, that he'll go home that very afternoon to the slice of meat he's just bought—and so by clinging to the image of his lunch he holds on, for a few hours, to the life he's been leading in Helsinki, returning in a pictured future to a place and a routine, a life he'll never actually resume.

Or if I could recover that feeling I had when I first learned algebra and geometry, subjects that didn't need to be taken in and memorized, the way you had to take in and retain other things. They were, it seemed, already right at hand. As though there were simply some latent machine I could turn on with logic, and then! An entire world I never suspected.

Two men on opposite sides of a desk, in a small, cold, dark room. André can't suppress his didactic habit, even under strain. The interrogator, off-blond and husky, proceeds through a series of questions, and because neither knows the other's native language, the two men speak in German. After the officer accuses André of having lied, André corrects his grammar.

Sie haben gelugt, says the officer.

Man sagt, "Sie haben gelogen," André replies.

The officer nods stiffly and continues. He produces a stack of suspicious documents the police have found in a search of André's apartment, including a long letter from Lev Pontryagin, a colleague in Leningrad who, because he'd been blinded in a childhood accident, dictated all his correspondence. And then it's as though the Russian's blindness invades the chamber where the interview is taking place. The city of Helsinki has declared a blackout, and the room grows darker and darker, until André can barely see his interrogator.

But who is this Pontryagin? the officer asks.

A mathematician.

And when did you first meet him?

At a conference in Brussels.

What was the date of the conference?

I don't recall the date. It was several years ago—August?

A young cadet arrives to paint the windowpanes blue, a job he must complete in that near-total darkness. He stands inches from the glass and feels his way with one hand while painting with the other. The officer asks nothing of significance while the young man is there. He lights a cigarette and, after some mumbling response from the cadet, stubs it out again. He asks about André's work, which André describes in more detail than he would normally share with a non-mathematician, hoping to convince the officer that he is only who he claims to be, even though he knows that being a French mathematician would not, in the other man's mind, disqualify him from being a Russian spy.

Once the window has been painted and the cadet leaves, the officer lights a gas lamp, hardly bright except by comparison with the preceding darkness. As their eyes adjust, both men drop their heads.

André spends the night in jail. The next morning, when he is escorted out, he thinks he's going to be executed. According to a story he would hear many years later, the chief of police did intend to kill him for being a spy, but the night before the planned execution, the chief attended a state dinner and there ran across a Finnish mathematician who knew André and who suggested he be deported instead. So he's put on a train, locked in a compartment with three other inmates, and delivered to the Swedish border, close to the Arctic Circle. There he is interrogated again, detained again. He has nothing with him but the clothes on his back, his passport, and his wallet. He is housed at a jail, with the local drunks, though he's free to go for walks and dine at a restaurant.

And then come a series of further journeys: by train to Stockholm and then on to Bergen, by ship from Bergen to Newcastle, this last leg a terribly rough crossing in an old behemoth that plows through a section of the North Sea that has been seeded with floating mines. André ignores the instructions to wear a life belt day and night, for he's sure he doesn't stand a chance of surviving a shipwreck in those waters in January.

The sea is calmer by the time the ship makes it to New-castle, but he's chilled by the sight of the harbor, so clotted with boats at anchor, ships that have been immobilized because of the fighting elsewhere, that it takes three days before his boat is allowed to dock. It's the first time that the war hits him so squarely in the face.

There's a certain kind of young kid for whom the word *algebra* has a magical shimmer, portending the enigmas of grades not yet reached, all the unimaginable revelations of junior high and high school. I remember this from around fourth or fifth or sixth grade, or whenever it was that I became aware that some of the eighth graders were learning an exotic math involving letters, and I see this even more strongly in my six-year-old son. Give me an algebra problem, he begs, and I'll comply, if somewhat reluctantly: $2x + 4 = 12$. Give me one with x's *and* y's, he'll say.

This isn't just him. One day a classmate comes over to play, and before I even notice that they've snuck into my office, they come marching back out, having dug up a treasure that my son's friend now presses to his chest: a raggedy old textbook. The royal-blue cover has worn down to cardboard at the corners, and the outermost layer of the spine has begun to peel away, exposing a strip of dried glue. The title, *Algebra*, appears in gold along the bowed binding and also on the front, which is otherwise undecorated except for the author's name and an image of a black rectangle with thin gold vertical stripes, as though algebra were a dark jail with golden bars.

They carry this book of mysteries, along with pencil and paper, out to the back porch, set everything down on the couch there, and, after briefly paging through the book and scribbling obliquely on the paper, proceed to climb all over the couch, up above the book and back down and then around it. They circle the couch with its totem as though in ritual worship. They might as well be constructing one of those Hindu altars, an invisible altar that is also an invisible math problem.

———

The word *cuneiform* comes from *cuneus*, the name for a wedge-shaped stamp used by the Babylonians to mark a clay tablet before it dried.

Dating from the second millennium B.C., the Babylonian "problem texts" reveal not only practical mathematics, calculations concerning inheritances for instance, but early stabs at algebra. The Babylonians used terms that expressed unknowns, and they posed and solved problems apart from any application. For example, "The *igibum* exceeded the *igum* by seven"—here *igibum* and *igum* mean reciprocals, n and $1/n$, though in other contexts they mean other things. As unknowns, *igibum* and *igum* may have lacked the economy of Descartes's x, but they indicate that even then, thousands of years ago, this practice of representation was under way.

I like the thudding sounds of these terms, in my invented pronunciation of them; to me they seem like bungling syllables trying to form words but not quite succeeding. I ought to put *igibum* on a plaque and hang it near my desk.

I'm acquainted with a pair of twins, one of whom is considerably larger, both taller and bigger, while the other is a more modest presence and slightly concave in the chest, which was presumably the result, they explained to me, of this smaller brother having been head-butted or jostled in the womb by the bigger one as he made room for himself.

Though Simone and André were born three years apart, they seem to me like those twins, yin-yang children who shaped and were shaped by each other. André was so sure of himself, while his younger sister was his first peer, but too young and too differently inclined to ever catch up; he could enjoy an older sibling's inflated belief in his own superiority while Simone, it seemed, was bent from birth by a sense of shame. That cavity in her chest, was he the first cause? Did her brother lean too hard against her ribs?

She once characterized friendship as "a miracle by which a person consents to view from a certain distance, and without coming any nearer, the very being who is as necessary to him as food," and maybe she was also describing the relationship of younger sister to older brother, the brother always leaping ahead and going away. The food she couldn't eat.

She believed that she had been poisoned, in her first year of life, by her mother's milk. "That's why I'm such a failure," she told people.

One evening my six-year-old wonders aloud whether some person long ago came up with the numbers, or whether they already existed before anybody thought about them.

People have been arguing about this for a long, long time! I say all too excitedly. Are numbers real or not? Were they discovered or invented? We pursue this question for a couple of minutes, by which I mean that my son thinks out loud with a logic I can't quite follow. Then we hit a wall. Neither of us knows how to take the matter any further. We move on.

At around this time my daughter, aged two, is learning to count by rote, so that part of our household soundtrack is her calling out numbers in the way that toddlers do, drawing out the vowels, her voice dropping into a lower pitch before ascending: *oooone, twoo-ooo, three-ee, seee-ven, elee-ven* . . .

A day or two later my son asks, "Can we talk about that thing again that we were talking about the other night? About 'Where are numbers?'"

Wanted for draft evasion, André is transported by ship to Le Havre, where an agent of British intelligence turns him over to French gendarmes. They take him to a dirty jail and lock him in a cell by himself, where he doesn't have access to math books or scratch paper. His only diversion is to peruse the obscene graffiti on the walls.

In a note he sends to his lawyer, he describes himself as at his wit's end. Alone in a cell, without the materials he needs to work, he says, there's no telling what he might do. In fact these histrionics are for the benefit of the prison administrators, who will read the postcard before mailing it; he hopes to scare them into treating him better. But his sister and his parents and his wife don't pick up on the ruse. They rush to Le Havre, only to learn that they won't be allowed to visit. They find the lawyer and haunt the parlor outside his office—which is stuffed full of red plush furniture, absolutely repulsive to Simone, like being inside the stomach of a pig.

She hectors the lawyer in her low flat voice to do *something*. She writes to an acquaintance in government: "You know what effect solitude, silence, and a lack of work can have on any human being; but my brother, especially, has never, I think, remained idle for an hour in his whole life, and he is accustomed to uninterrupted intellectual activity."

In other words, a war has begun without altering the Weil family belief that the genius son must be allowed to continue his work, without interruption, never mind the fact that he's in prison. Never mind that soon enough they'll all need to solve a more pressing problem than any in mathematics, namely, how to get out of France. Then again, maybe the Weils' belief in the importance of André's work is only a more extreme variation, distributed among family members, of the wool that any would-be intellectual or artist must keep pulled over her eyes, shutting out enough of the world and its strife to stay at least somewhat convinced that her work could be worth doing.

With the help of one of the guards, André sends a message to his family that he's not as unhinged as the postcard might have implied. The prison officials move him into the neighboring cell, with two other men: one who was caught with the dead quail he'd poached from a private estate, another who doesn't reveal his offense but takes from his pocket a tarnished sou he's managed to keep and starts to trace new graffiti into the wall. André looks on in envy, sitting on his hands, dying to grab the coin and write something himself.

6. NOT LONG AGO, I DISCOVERED THAT all the lectures from a certain Harvard math class are on YouTube. These were filmed, in 2003, so that they could be watched by online viewers as well as by the enrolled students. (An early foray into e-learning, it seems.) The class was an iteration of one that I'd taken in 1993, a one-semester course in abstract algebra, and the textbook was that same book my son and his friend had dug out of my office, though a different person was teaching. My class had been taught by a beautiful Italian woman, a visiting professor—I remember hearing rumors that at least one male mathematician in the department had lately taken up the study of Italian. Her lectures had been clear and well punctuated and imbued with a kind of wholehearted gravity, and as she stood there laying out some theorem about symmetry groups or Euclidean domains, I used to marvel at her, wondering what this woman would have become had she been born fifty years earlier. A bookkeeper in some Italian store?

I started to watch these online lectures and to half-ass the problems assigned as homework, looking to recapture some of what I'd been occupied with in college. But when you're just out of high school, chances are you've studied some sort of math almost every year that you can remember being alive, and so even the more abstract realms of mathematics seem connected to something you're used to doing. Twenty-some years later, twenty-some years in which I'd become a writer and rarely thought about abstract math or had even a glancing encounter with abstract math, the stuff of university-level algebra seemed very, very remote.

And still, it was beautiful. I'm ambivalent about expressing it that way—"beauty" in math and science is something people tend to honor rather vaguely and pompously—instead maybe I should say that still, it was very cool. (This is something the course's professor, Benedict Gross, might say himself, upon completing a proof: "Cool? Very cool.") A quality of both good literature and good mathematics is that they may lead you to a result that is wholly surprising yet seems inevitable once you've been shown the way, so that—aha!—you become newly aware of connections you didn't see before.

Yet in math these surprises break loose from their creators. They find a place in the firmament of what's been discovered. To me, one thing math always had going for it was its solidity; its theorems seemed not only ingenious but true. They were facts of the universe. As a teenager I always felt the ground moving under my feet, and there was something fixed and unassailable about math. At the same time, math seemed to be a domain unto itself, at a comfortable distance from daily life. It was as though there were another world besides this one—private, constructed mentally, a symbol world—and yet it turned out to house a map of this one, in some sense it *was* this one too.

"What's the ontology of mathematical things? How do they exist?" the mathematician John Conway once said. "There's no doubt that they do exist but you can't poke and prod them except by thinking about them. It's quite astonishing, and I still don't understand it, despite having been a mathematician all my life. How can things be there without actually being there?"

Everything is number, according to the Pythagoreans, a secret brotherhood in the sixth century B.C. that wove mysticism into mathematics and vice versa, cultivating ideas of order that had begun with the scratching of figures into the sand, the press of a wedge into clay.

Pythagoras himself is a hazy figure, his story and his teachings recorded only elliptically by his contemporaries. Born around 570 B.C. on the island of Samos, he traveled widely, consulting with Egyptian priests and Babylonian architects before settling down in Croton, in southern Italy, and founding a school. He and his followers subscribed to a doctrine they kept secret, one that seems to have combined mathematics and religious teaching and to have contained in embryo the concept that the objects of mathematics are mental objects, abstractions—but abstractions believed to be in intimate correspondence with the essence of the universe. The Pythagoreans thought that the only numbers were the whole numbers (1, 2, 3 . . .) and their ratios.

They also thought that numbers were friendly, perfect, sacred, lucky, or evil.

Ultimately their theory fell apart. They realized that if you have a square in which the sides measure, say, one unit, the length of its diagonals cannot be expressed as a ratio of whole numbers. There had to be other kinds of numbers, what they called incommensurables and we know as irrational numbers.

Supposedly the cult tried to suppress this truth. Meanwhile, agitators in Croton attacked the Pythagoreans, and, as different sources have it, Pythagoras himself perished in a fire, or died of starvation, or committed suicide, or was murdered.

Some more number types of the ancient world: perfect, excessive, defective, and amicable. Another story has it that one Hippasus of Metapontum discovered incommensurables, and that as a result he was thrown off a ship and drowned.

Professor Gross, that is to say the 2003 version of Gross preserved on YouTube, has gone half gray, meaning his beard is gray but the hair on his head and eyebrows remains dark. He is no longer young, not yet old. Every so often he strikes an avuncular note, reassuring the students that this material might seem hard at first but they'll get it, they'll do fine. I'm fond of him, the miniature mathematician inside my laptop.

One day, he turns and looks straight at the camera. "Hi, people online!" he calls out.

That's me! Hi!

What we know as the Pythagorean theorem—for a right triangle, the square of the length of the longest side is the sum of the squares of the shorter sides' lengths—was discovered well before Pythagoras. Hundreds of proofs of the theorem have been devised over the centuries, one of them by James A. Garfield, who was a member of Congress when his proof was published in 1876 and who five years later would become the first left-handed U.S. president. And who died later that same year, after having been shot by Charles J. Guiteau.

André is transferred from Le Havre to a military prison at Rouen, where he is allowed paper and pencil and books. *My dear sister*, André writes in February. "I received your letters, the one that went to Le Havre and the one that you sent me here. I'm angry that you can't see me right now. But I don't think that will last." He tells her that he's been spending the daylight hours correcting a draft of an article, an exacting and mechanical labor that serves him well after so long without work. It's as though he's relighting his mind, one candle at a time. He occasionally takes a break and reads a novel, and he stops for the day at sundown, since there's nothing but a window to read by.

His cell is long and narrow, its consolation a small writing table attached to the wall. On one side of the cell is the window, high and barred; on the opposite side is a thick, heavy door with a circular peephole. Every so often, when he happens to glance up from his work and look in that direction, André sees some portion of a guard's face pressed right up to the hole, staring at him.

Finally he is permitted to see visitors twice a week, though no more than two at a time. Because there are four Weils

(Bernard, Selma, Eveline, Simone) who come by train, sometimes with Alain, Eveline's son by her first marriage, one or
two of them will go to visit André while the others remain
at the station, sitting on the wooden benches of the waiting
room there. The Weils are an anxious and bumptious and hy-
perintelligent family unit, with an artillery of wicker baskets and packages and sausages and books. The five of them,
muttering and worrying and occasionally reciting a line of
poetry, come and go from that waiting room like birds to
and from a tree. One pair flies off, then returns; another pair
departs.

At visits Simone is told where to stand, separated from
her brother by two iron grilles, between them a passage where
a prison guard marches back and forth. She calls to André
in Greek, and the guard barks that they must speak only
French. They trade off, a halting chant of practical concerns
and empty but heartfelt assurances. The sister, holding tightly
to the bars, has the same eager face she had at age three whenever her six-year-old brother invited her to play with him.
The brother, paler than usual but in decent spirits, says that
he has instructed the editor of a certain journal to send page
proofs of his article to her so that she can copyedit them. In
Greek, forgetting herself, she says she'll do it gladly. The
guard informs them the visit is over.

They carry on a separate, deeper dialogue in the letters
they send back and forth during this period, which are long
and cerebral and discuss at length the mathematics of the
Babylonians and the Pythagoreans. It's a throwback to the
intellectual closeness they had as children, the older brother
once again playing tutor to the younger sister—but now her

devotion has a sharper edge to it. She'll keep dogging him to tell her about his research. She'll ask, What is the value of work so abstract and specialized that it has no meaning to the common person?

Certain later Pythagoreans believed that the human soul is reincarnated every 216 years. Probably they favored the number 216 because it's the cube of 6—but then why the cube of 6?

Embedded within the shaggy narrative of Don DeLillo's 1976 novel *Ratner's Star*—about scientists in a trippy '70s sort of future—are several wonderful depictions of mathematical reasoning. About a mathematician solving a problem, DeLillo writes: "He scribbled calmly, oblivious to everything but one emerging thought, feeling the idea *unerase* itself, most evident of notions, an idea with a history . . . What breathless ease, to fall through oneself." I love that coinage of "unerase" to describe how something created might seem already to have existed. And the way that thought and feeling here combine in a spell of self-forgetting. Again breathlessness. Again I wonder, was it the truth André Weil was after, or this feeling? Maybe it doesn't matter, since they go hand in hand.

Though it was not the same thing as making a new discovery, the closest I ever came to such a feeling was while working on a homework problem from that algebra class I took in college. I don't remember the specific problem but I remember being stumped for a long time, trying versions of the same doomed attack and not getting anywhere, and then, while I was doing something else, it came to me that I could get at the answer by constructing an entity not given in the original problem, a family of functions that behaved in a certain way, which would form a kind of bridge to a solution. I didn't trust myself, because it seemed that I had just arbitrarily made something up, but the method seemed to work. It was the middle of the night and I was delighted. The next day I ran into a guy also taking the class, who was small and soft-spoken and from Hungary, a country that seems to export mathematicians as one of its principal products. He was ahead of me in math generally, but he hadn't figured out the problem yet, and I showed him my solution.

Nice, he said, nodding.

That was probably the high point of my mathematical career.

Simone bides her time in that train station waiting room, her face eclipsed behind glasses and hair, her body draped in heavy clothes, her spirit likewise cloaked in heavy intellectual armor, in argument, in the whole weight of Western civilization. Let's say she's writing now, while sitting on a hard bench with her books strewn around her. Scribbling in anger. She wishes she were the one suffering in jail, and somehow this makes her regret even more the fact that her brother's great passion is something she can't understand, though in her letters she tempers her frustration. She writes that she hopes to be able to visit him soon, given that it's impossible for them to change places as she would prefer.

In the meantime she proposes to him that he try to clarify the nature of his mathematical research. Could he explain it to her, since after all he has some extra time on his hands? She would like to know "what exactly is the interest and significance of your work." Even if ultimately there's no way to convey it fully, she says, he might benefit from the effort, and she would surely find it interesting.

"Inasmuch as I am less interested in mathematics than in mathematicians," she writes.

Her own capacity for work has lately been low, she says, and so she's undertaken what she considers light duty: learning the language of the Babylonians. She studies from a bilingual text, and also reads the *Epic of Gilgamesh*, with its story of a friendship cut short by death.

In her next letter, she again presses him to explain what he's doing, claiming that she can't remember whether she's brought this up already. "What would be the risk?" she writes. "You don't risk losing time, since you have time to lose." Maybe there's a way to account for what you do, to make it clear to nonspecialists, she suggests. "This makes me passionate."

The trains come and go, heaving their way back to Paris. She pauses to help Alain with his ancient Greek.

She has managed to get her hands on a book by Otto Neugebauer, a scholar who's undertaken the labor of translating the mathematics contained in cuneiform tablets into German. Simone copies one of Neugebauer's Babylonian algebra problems in her letter to André. The problem gives the dimensions of a canal to be dug, the amount of dirt that a worker can dig per day, and finally the sum (but only the sum) of the number of days worked plus the number of workers. The problem requires the solver to find the two numbers that make up that sum: how many workers, how many days worked. "Funny people, these Babylonians," Simone writes, for it's a ridiculous calculation. In no actual canal-digging scenario would something like the sum of days worked plus workers be known, without knowing the two quantities individually—and this, to her, speaks to the Babylonian way of thinking. Such a problem is only abstract, a manipulation without any reality behind it.

"Me, I don't so much like this spirit of abstraction," she writes. "But you must be descended directly from the Babylonians." In her letters Simone has a bone to pick with algebra and with abstraction itself; for her, there is something distasteful about mathematical thinking untethered from any study of nature. This type of math, in her eyes, is merely a game, referring only to itself. She prefers geometry, by which she means the geometry of the ancient Greeks. ("I think that God, as the Pythagoreans said, is a geometer—but not an algebraist," she writes.) Given that the Pythagoreans and their successors surely would've known about Babylonian algebra but didn't incorporate it into their work until centuries later, she infers that her beloved Greeks objected to algebra just as she herself does. She goes so far as to suggest that there must have been a religious injunction that caused them to steer clear.

As in her thesis about Descartes, Simone interprets history in a way that seems eccentric, if not outright bonkers. And this animosity toward algebra—where did it come from? Was it that she resented her brother for leaving her behind, as he entered into his ethereal vocation? Was it something that stewed in her as she wandered around those Bourbaki conferences, listening to grown men speaking in jargon she had no way of understanding?

But I don't think it was quite that, since even as she questioned her brother about his work, she praised him and saw his research as very valuable—in other letters she urged him to keep it up, and when he was jailed in Le Havre, no one was more frantic than she to get him his materials. My guess is that while her distrust of abstraction can't be entirely separate from her relationship to her brother, it's more directly tied to her search for meaning. Everywhere people were suffering, under threat. Mathematicians, she felt, should continue their work regardless, but at the same time the work, all intellectual work, should tell us something true about the world, shouldn't it?

"After this his fame grew great," the ancient Roman philosopher Porphyry of Tyre wrote of Pythagoras, "and he won many followers from the city itself (not only men but women also, one of whom, Theano, became very well known too) and many princes and chieftains from the barbarian territory around. What he said to his associates, nobody can say for certain; for silence with them was of no ordinary kind."

Neugebauer, the scholar of ancient Babylonia, was an Austrian who fought in World War I and wound up in the same prisoner-of-war camp as Ludwig Wittgenstein. After Hitler rose to power, Neugebauer found a job in Denmark and over the course of several years published three volumes of his translations. Then, in 1939, he emigrated to America, where in collaboration with another scholar he published an English version of his magnum opus, as *Mathematical Cuneiform Texts*.

He would also write and publish a journal article called "The Study of Wretched Subjects."

In his lectures, the Professor Gross of 2003 manages to impart something like suspense and drama to a subject without much inherent narrative tension. There's an urgency to his presentation, a vigor born of logic itself. "Now I claim," he says, raising his voice and drawing out the *I* and the *claim*, like a magician announcing his next trick—"Now *I* CLAIM . . . that $r = 0$."

He has a fondness for historical digressions (as do I, obviously), and during the first week of class, he presents one as though it were a theorem: "*If* you go to Paris and *if* you take the *ligne de sud* to the Parisian suburb of Bourg-la-Reine, an *absolutely disgusting* suburb, and you go to the main intersection of Bourg-la-Reine, an *absolutely disgusting* intersection, you will find a plaque that says, *Ici est née Évariste Galois, mathematicien*."

And *if* you give an introductory algebra course, surely you *must* mention Galois. A misunderstood, mournful prodigy, his short career a series of arguments and rebuffs, he bequeathed to the world not only the fundamentals of what is called group theory but also the romantic image of the mathematician in its purest, most distilled form. He died in 1832, when he was twenty, as the result of a duel: pistols at dawn, the whole shebang. His father, a provincial mayor slandered by enemies, had committed suicide. The papers he'd submitted to the Académie Française had been lost or rejected. He'd been jailed for making treasonous remarks while drunk at a dinner. The night before the duel, or so the legend goes, he tried to write down everything he knew but hadn't yet committed to paper, all the math swirling around in his head, pausing to scribble in the margins, "I have not time, I have not time."

This tale circulated for many years, though it was steeped in exaggeration. (For instance, Galois wrote up his major mathematical ideas in those papers he submitted to the Académie Française, rather than on the eve of the duel.) Galois's story aside, his insights helped transform the study of equations, which in the nineteenth century would be flipped on its head. He and others developed an inspired new approach: Instead of trying to understand an equation by finding the solutions, one could presume the solutions and examine the domain in which those solutions exist. To say it another way, given the polynomial equations we learned about in school (equations like $x^2 - 5x + 6 = 0$) and were tasked with finding the roots of (here $x = 2$ and $x = 3$), you could take a step back into a more abstract realm: instead of solving for those roots, you could assume they existed, designate them by symbols (u and v rather than 2 and 3), and consider the smallest set containing u, v, and all the numbers you could obtain from them by adding, subtracting, multiplying, and/or dividing. By investigating the properties of such sets of numbers, rather than the original polynomials, you could cast the study of equations in a new light.

"Gradually the attachment of the symbols to the world of numbers loosened, and they began to drift free, taking on lives of their own," the historian and novelist John Derbyshire writes about this shift in algebraic thinking. What the x might represent mattered less than what the x might do in various circumstances. It was the sort of turn toward abstraction that Simone would come to hate but was hugely fruitful in a way that she would refuse to see.

When the Pythagoreans claimed that the universe was number, they meant it, literally; for them creation had begun with a unit, a number somehow possessed of spatial magnitude. First there was a single unit, then it split into two units, and as it did so, the universe drew into itself a void that would keep the two units apart—as though taking in breath, they said. The universe inhaling the void. The two units begat the first line, and then, by more splitting, a triangle, then a tetrahedron. From a single point all of geometry spilled forth, and this geometry constituted the material world.

Much of this belief system was noted by Aristotle, who wrote about Pythagorean thought and, more often than not, dismissed it as absurd. "All . . . suppose number to consist of abstract units, except the Pythagoreans; but *they* suppose the numbers to have magnitude."

Every day André is permitted a thirty-minute walk inside the circular prison yard. The yard is divided into sectors, and the prisoner is restricted to one of them, as the guards watch him from a tower in the center. Though André normally wears sandals even in winter, here he has only the ill-fitting, bulky work shoes that he was given in Finland. Nonetheless he tries to walk fast, clopping along in his brogans, around and around the sector's edge. When he walks toward the guard tower he thinks about complex functions. When he walks away from it he cranes his neck to see the birds, the clouds beyond the fence.

"My dear sister," he writes, when he is back in his cell, "Telling nonspecialists of my research or of any other mathematical research, it seems to me, is like explaining a symphony to a deaf person. It could be attempted, you could talk of images and themes, of sad harmonies or triumphant dissonances, but in the end what would you have?

"A kind of poem, good or bad, unrelated to the thing it pretends to describe."

You might compare mathematics to an art, he goes on, to a type of sculpture in a very hard, resistant material. The grains and countergrains of the material, its very essence, limit the mathematician in a manner that gives his work the aspect of objectivity. But just like any work of art, it is inexplicable: the work itself is its explanation.

As for Babylonian algebra, it did in fact infiltrate Greek mathematics, but it was algebra translated into geometric terms, he writes. Take for example the work of Apollonius, in which algebraic equations became conic sections: parabolas, hyperbolas, ellipses. Yet the most original thing about Greek mathematics is that the Greeks didn't deal in approximations; they killed number for the benefit of Logos. In other words the Greeks, rather than just using numbers to calculate, considered them as pure quantities, and this abstraction—that is, this conceiving of the whole numbers as such—constituted the leap forward from which everything else followed.

So here we have André in prison, Simone at loose ends, all Europe going to hell—and they take to arguing about the nature of ancient mathematics.

(But who's to say, when the world is going to hell, that you shouldn't argue about ancient mathematics?)

When it was discovered that the whole numbers couldn't fully account for even simple geometric relationships, the Greeks had to start over, he writes, at the foot of the hill. Since one could no longer be sure of anything.

Another morsel from *Eros the Bittersweet*: our word *symbol* comes from the ancient Greek *symbolon*, which was half of a knucklebone, "carried as a token of identity to someone who has the other half," like one of those cheap heart-shaped lockets that break into two pieces. I imagine André and Simone as having such half knucklebones. Each sibling carrying one, as a symbol.

Although I've never been susceptible to phantom sense impressions, one day as I watched an algebra lecture on my laptop I began to smell chalk dust so strongly that I wondered whether my kids hadn't stashed some sidewalk chalk in my office. (They hadn't.) Thick in my nostrils, I could've sworn: vintage chalk dust from 2003.

That year, the class reminds me, was a long time ago. In one lecture, a phone starts to ring, and Professor Gross, with a very 2003 self-consciousness about the technology, pulls out a flip phone, opens it, and without preamble asks, "Is the governor of Puerto Rico here?"

In 2003 he is also the dean of Harvard College, and as he explains after the call, he is supposed to receive the visiting governor of Puerto Rico later that day.

He turns to face the class. "Moocho goosto!" he says with relish—*mucho gusto*, pleased to meet you. Then: "I'm practicing my Spanish."

In a postscript to that letter in which André tells his sister that it would be pointless to explain his work, he relates a story about the nineteenth-century Norwegian mathematician Sophus Lie—for whom I myself have a sympathetic fondness, because he was a tall Scandinavian who for a long time couldn't make up his mind about what to study or which career to pursue. As a university student, Lie had first tried his hand at languages, then switched to science, after failing Greek. He would later say that he found the road to mathematics long and difficult. But he had a thing for long roads—one weekend he walked the sixty kilometers from the capital to the town where his parents lived, only to find that they were not at home, at which point he turned around and walked back—and when he was twenty-six, he encountered certain new ideas in geometry that led him to embrace math for good.

"The powerful Northman with the frank open glance and

the merry laugh, a hero in whose presence the common and the mean could not venture to show themselves," wrote Anna Klein about Lie. Anna was the wife of Felix Klein, another renowned mathematician and a friend of Lie's. As young men they both lived in Paris, in adjacent rooms. Felix Klein would recall that on one morning he had risen early and was headed out for the day when Lie, still in bed, called to him. During the night, Lie explained, he had discovered a connection between lines of curvature and minimal lines.

I did not understand one word, Klein would write. Yet later on, after he had left and was going about his business, he perceived in a flash what Lie had meant and how to prove it geometrically. That afternoon he returned home and, finding that Lie had gone out, wrote a letter containing the proof.

I imagine the excitement he must've felt as he slid it under Lie's door.

Dude, that is so rad.

In another of his lectures, Gross pauses to consider a particular function (if you're curious: the function $f(t) = e^{2\pi it}$, along with its derivative $f' = 2\pi if$), and then he invokes Lie.

"This is what Sophus Lie discovered," Gross says, with feeling, "that the study of group homomorphisms in the context of continuous groups is intimately related to the solution of differential equations!"

Leaving aside the details here, the gist of the matter is that Lie, by expanding a theory in algebra, namely the study of groups, found a way to shed light on an entirely separate area of math—or one that had seemed entirely separate, namely differential equations. It's as though he located a wormhole from one mathematical realm to another.

"Is that amazing?" Gross exclaims. A pause, and then: "But I digress."

The Franco-Prussian War forced Klein and Lie to flee Paris. Lie, naturally, left on foot. Near the town of Fontainebleau, he was arrested and (as André wrote to Simone) accused of spying for the Prussians. His mathematical notes were thought to contain coded military secrets, just as André's papers would draw the attention of the Finnish police seventy years later.

"Occupied unceasingly with the ideas which were fermenting in his head, he walked in the forest each day, stopping at the places furthest from the beaten track, taking notes, drawing diagrams in pencil," recalled one of his French colleagues, Gaston Darboux. "At that time this was quite enough to awaken suspicions." Darboux appealed to the imperial prosecutor, insisting that the notes had nothing to do with national security, and after a month in prison, Lie was released.

He returned to Norway, though in his native country he eventually came to feel isolated, ignored by mathematicians working on the Continent. In a letter to Klein he complained that Darboux "plunders my work"—that is to say, the Frenchman who'd pleaded to save Lie's neck was stealing his results and publishing them as his own.

———

Simone imagines that a young paratrooper has landed on the terrace of the Weil family apartment, a bewildered German teenager for whom, in her reverie, she feels nothing but tenderness. At dinner she asks her parents whether they would offer shelter to a German soldier.

Absolutely not, says her father. I would turn him in, of course.

She refuses to eat another bite until he promises her that he would help the young man, this hypothetical soldier she saw in a daydream.

Selma Weil, who has always fretted about her daughter, now worries about her son, too, not to mention the fate of France itself. She goes all around the neighborhood sharing gossip and advice, buying up more food than her kitchen can hold, huffing her way upstairs and back down again, and if she happens to sit still, she has the twitchy look of a hare. Her dark eyes are anxious, vigilant. She's had her bangs cut very short, which has the contradictory effect of making her look girlish while exposing her lined forehead. One day she absently pulls a button off her sweater, unaware that she's been tugging at it until she finds it in her hand. She stares at it in surprise, then mutters to herself in Russian, much as she claims to have forgotten every word of the language she spoke as a young girl. She writes to cousins in California who have invited the family to come live with them. She rereads André's letters. She plays her beloved Beethoven sonatas, too loud and too fast.

————

"What can I say about myself?" André writes to Eveline. "I am like the snail, I have withdrawn inside my shell; almost nothing can get through it, in either direction."

Only paper, that of his correspondence, his manuscripts, his books. Because he is allowed those things, it is a tolerable existence, that of a snail who executes the following actions: work, eat, write, read, sleep. While his life is stripped-down and constrained, he strides ahead in his mathematical investigations, making progress in the area of algebraic curves.

His colleagues, who'd previously written him to offer their sympathy, now envy him instead, reminding him that not everyone is so lucky to sit and work undisturbed. "I'm beginning to think that nothing is more conducive to the abstract sciences than prison," he tells Eveline. Her letters back to him are reassuringly full of activity; she says that she's been cultivating pea plants, trying to teach Alain about fractions. In a sense this arrangement merely heightens a division of labor they'd already adopted, André off by himself, contemplating mathematics, and Eveline out in the world, with her child and later with their children.

"A heavy, opaque, suffocating atmosphere has settled over the country," Simone writes to André. "People are downhearted, discontent, but they also tend to swallow whatever they are served without protest or surprise. Characteristic situation of a period of tyranny. Unhappiness is joined to an absence of hope. France is and will be for a long time (unless there's a social convulsion) in a state of torpor and resignation."

She sends a short letter while working on a draft of a longer one, funneling a fury of thoughts into her neat, upright script. She sets the finished pages next to her on the train-station bench and tucks the edges under her thigh while she prosecutes her argument about the ancients. Countering her brother, she insists that the Greek stance toward algebra can't be explained merely by saying that they assimilated algebra into their geometry. It must have been taboo—algebra must have seemed impious, she writes. Mathematics, for the Greeks, was not just a mental exercise but a key to nature. It illuminated a structural identity between the human mind and the universe.

As for André's comparison of mathematics to art in a hard material, Simone has her doubts: What material? The proper arts work on material that exists physically. Even poetry has the material of language regarded as an ensemble of sounds. The material of mathematical art is a metaphor, and to what does it correspond?

The material of Greek geometry was space, but space in three dimensions, actual space, a constraint imposed on all men. It is no longer this way. Your material, she says to André, isn't it just the ensemble of previous mathematical work, a system of signs? Rather than a point of contact between man and the universe, Simone argues, present-day mathematics has become inaccessible.

———

In 1886 Lie moved to Leipzig to take over a position from Klein, who'd been hired by the University of Göttingen. There he became known for his informal way of teaching (Felix Hausdorff was one of his students), yet he began to chafe at the workload and the fact that his colleagues treated him as Klein's disciple. He found it harder and harder to sleep at night, and finally he suffered a nervous breakdown. At a psychiatric hospital near Hanover, he resisted the prescribed opium treatment. Although he returned to teaching the following year, it took that year and the next before his insomnia passed. "With sleep," he wrote, "the pleasure of life and work has returned."

But he was a changed person, touchy and irritable. He wrote to friends in Norway that he longed to return there. He clashed with Klein, his friend of more than twenty years, then publicly attacked him. ("I am no pupil of Klein, nor is the opposite the case, although that might be nearer to the truth," he wrote.) Klein found this "both painful and incomprehensible," Anna would recall, but "soon my husband understood that his best friend was ill and could not be held responsible for his acts."

A few years passed, and then one summer evening, as Felix and Anna were returning home from an excursion, they found their estranged friend waiting for them. "There, in front of our door, sat the pale sick man," Anna wrote. "'Lie!' we cried, in joyful surprise."

Felix Klein and Sophus Lie "shook hands, looked into one another's eyes, all that had passed since their last meeting was forgotten."

Lie went back to Norway in 1898, but he could only lecture for a few months before his deteriorating health forced him to retire. He died, of pernicious anemia, the following year.

At last André goes on the offensive, that is to say, he answers Simone's repeated requests with a long, technical description of some of his mathematical work, a treatise in the form of a letter. He knows full well that she won't understand these "thoughts," as he calls them: "I decided to write them down, even if for the most part they are incomprehensible to you." He plunges into a density of terms she wouldn't know, with only minimal efforts to say what he means by quadratic residues, nth roots of unity, extension fields, elliptic functions.

In the first half of his letter he sketches a historical context for his work, starting with the nineteenth-century watershed in algebra, that leap by which mathematicians inverted the problem of solving equations within given domains by constructing domains in which given equations had solutions. He alludes to a time when questions about numbers began to rub up against questions about equations or functions in new ways. "Around 1820, mathematicians (Gauss, Abel, Galois, Jacobi) permitted themselves, with anguish and delight, to be guided by the analogy between the division of the circle . . . and the division of elliptic functions," he writes.

Anguish and delight! As he's laying out his none too explanatory explanation of his research, André emphasizes the role of analogy in mathematics—which his sister might appreciate, even if the rest of it flies right over her head. Here, analogy is not merely cerebral. The hunch of a connection between two different theories is something felt, a shiver of intuition. For as long as the connection is suspected but not entirely clear, the two theories engage in a kind of passionate courtship, characterized by "their conflicts and their delicious reciprocal reflections, their furtive caresses, their inexplicable quarrels," he writes. "Nothing is more fecund than these slightly adulterous relationships."

Analogy becomes a version of eros, a glimpse that sparks desire. "Intuition makes much of it; I mean by this the faculty of seeing a connection between things that in appearance are completely different; it does not fail to lead us astray quite often." This, of course, describes more than mathematics; it expresses an aspect of thinking itself—how creative thought rests on the making of unlikely connections. The flash of insight, how often it leads us off course, and still we chase after it.

Pernicious anemia, the cause of Lie's death, is a decrease in normal red blood cells that results when the intestines cannot absorb enough vitamin B_{12}. Symptoms include confusion, depression, loss of balance, and numbness of the hands and feet.

David Hilbert, one of the leading mathematicians of the early twentieth century, also suffered from pernicious anemia, and although he benefited from a new treatment developed in the 1920s, the disease contributed to his decline at the same time as his beloved Mathematical Institute in Göttingen, for decades a hothouse of mathematical progress, was drained of its talent by the Nazis. One evening Hilbert attended a banquet where he was seated next to the Nazi minister of education, who asked him whether it was true that the institute had suffered following the removal of its Jewish faculty and their supporters. "It hasn't suffered, Herr Minister," Hilbert replied. "It just doesn't exist anymore."

Dementia set in, and he came to believe he was living again in Königsberg, the Prussian city of his childhood. He died in 1943; hardly anyone went to his funeral.

But I digress.

Simone studies her brother's letter closely, so closely that it hurts. She pulls some of his books down from their shelves and begins to wade through a German text on complex functions, until her head threatens to split open.

Yet might she be right, or at least not altogether wrong, to think her brother's work should be explicable? That it should do more than just extend the work that preceded it, that it should reveal something about the world?

Meanwhile André sends her another letter, two days later, continuing their debate about ancient mathematics. In their correspondence both he and Simone propose ways in which the Greeks might've discovered irrational numbers, drawing geometrical diagrams in the margins to illustrate their theories. Did the discovery trouble the Greeks or inspire them? André suggests that they were disturbed by it, while Simone counters that it would've brought them joy.

In one, two, three drafts that she composes in reply to the second letter, again there are figures traced in the margins, as well as more geometrical ideas, discussions of Platonic dialogues, musings on the relationship between an artist's worldview and her art, speculations about mysticism in ancient Greece.

Simone conceives of a civilization in which mathematical reasoning, mystical belief, and existential loneliness formed an energetic triangle. The Greeks, she writes, experienced intensely the feeling that the soul is in exile: exiled in time and space. Mathematics could bring some ease to the exiled soul, she says. Doing math could free you from the effects of time, and your soul could come to feel almost at home in its place of exile.

She also writes, regarding André's explanation of his research, "I understood nothing of your sixteen-page letter (which I read several times)." She has nonetheless sent excerpts to his lawyer, thinking perhaps a description of his research might be useful for his court case. And she tells him, "Send me more things like that if the inspiration strikes. I like it very much."

It wasn't lost on me, as I watched math lectures on my computer, that my return to (casually) studying math, much as it was not quite intentional, much as I stumbled into it, had come to resemble one of those self-improvement projects that middle-aged people embark on, like learning a language or taking piano lessons or rereading the classics. And yet studying German or piano or Homer seemed downright practical, relative to Math E-222.

What lies behind these efforts? Nostalgia? Wanting to stave off the inevitable, already perceptible decline? I suppose both, and then of course (almost too obvious to mention, and I feel like a scold and a cliché for even bringing it up) there is simply the enterprise of attention itself, the goal of concentrating on anything at all, in this era of distraction technologies. Surely Simone Weil, as odd as some of her beliefs and proposals were, was right to emphasize the importance of sustained attention, which is something we are letting slip away, or really giving away, with little more than mild, fleeting second thoughts.

Great titles in math:

> *The Whetstone of Witte*
> *What Are and What Should Be the Numbers?*
> *New Invention in Algebra: As Much for the Solution of Equations as for Recognizing the Number of Solutions That They Have with Several Other Things Necessary for the Perfection of This Divine Science*
> *The Sand-Reckoner*
> *Logarithmatechnica*
> *The Mathematical Science Reduced to Tables*
> *Essay on Fire*

———

Simone understands more of André's disquisition than she lets on. She includes a detailed summary of it in her letter to his lawyer, presented in the unlikely hope that the military tribunal could be convinced of the importance of André's work and that her brother would thus be freed.

I picture this lawyer as a harried sad sack, exhaling as he pulls from its envelope an overly long missive sent by the same pestering angular female who has been unrelenting in her visits to his office. He reads the opening lines: "Number theory (putting aside some propositions discovered by the Greeks, notably the Pythagoreans) begins in France, in the seventeenth century, with Fermat. It was then the German mathematician Gauss, in the beginning of the nineteenth century, who made the most decisive progress . . ."

He blinks and sets the letter down on his desk.

Sophie Germain, born in 1776 in Paris, could not, as a woman of that era, be admitted to the École Normale or the École Polytechnique, but she studied mathematics as best she was able, borrowing from other students' lecture notes and studying the books in her father's library. She wrote letters to the great mathematicians of her day under a pseudonym, M. Le Blanc.

"But how can I describe my astonishment and admiration on seeing my esteemed correspondent Monsieur Le Blanc metamorphosed into this celebrated person," replied Carl Friedrich Gauss, after she was compelled to reveal her true identity. By then she had withdrawn from social life, weary of being regarded as a curiosity because of her mathematical talent. "The taste for the . . . mysteries of numbers is very rare; this is not surprising, since the charms of the sublime science, in all their beauty, reveal themselves only to those who have the courage to fathom them. But when a woman, because of her sex, our customs and prejudices, encounters infinitely more obstacles than men, in familiarizing herself with knotty problems, yet overcomes these fetters and penetrates that which is most hidden, she doubtless has the most noble courage, extraordinary talent, and superior genius."

He challenged her to come up with her own proofs for three of his latest theorems, which she duly sent back to him. Six months later, he sent his last letter to her. He had accepted a professorship at Göttingen and, he explained, had become too busy to keep up their correspondence.

"Remain always happy, my dear friend," Gauss wrote to Germain before terminating their exchange for good. "The rare qualities of your heart and mind deserve it, and continue from time to time to renew the gentle assurance that I may count myself among your friends, a title of which I will always be proud."

At trial, André is defended poorly by his lawyer and sentenced to five years in prison, and so he agrees to serve in a combat squad, in return for a suspended sentence. One ordeal ends and another begins; he joins a machine-gun unit, alternately lifting chests full of ammunition and sneaking off to read math books. During a German offensive, the men are evacuated to England, where, because he misses curfew one evening, André is put in "prison," that is to say he's confined to an area of tents surrounded by barbed wire. Separated from the rest of his company, he literally misses the boat when the others are sent to Morocco. He becomes, instead, an interpreter for the guards who've been his jailers, and as such enjoys a good deal of spare time, free to go to the library and visit colleagues at the University of Bristol. He's managed to turn military service into a rather leisurely, math-centered existence.

The Germans, meanwhile, have reached Paris. Simone and Bernard and Selma squeeze onto the last southbound train and arrive in Nevers, in central France, barely ahead of the occupiers. They continue on foot toward Vichy, buying themselves baskets in an effort to pass as peasants, though this only makes them look like three well-off urban Jews carrying baskets.

André makes his way to London, where he flirts with English women and persuades a friend in the camp office to fill out a card stating that he has pneumonia. A hospital ship takes him to Marseilles, and he reunites with Eveline and Alain. The three of them board another ship, sailing across the Atlantic to Martinique, and from there he secures an American teaching position and a visa.

His parents and sister remain in the south of France, in Vichy and then in Marseilles, trying to obtain visas to enter Morocco or Portugal, with the idea that from there they'll find a way to follow André to the United States.

PART THREE

7. IN A DREAM, ANDRÉ MEETS JACQUES Hadamard and notices that his beloved old teacher is wearing an undershirt and short pants and that as a matter of fact André is wearing the same thing himself. The kindest man he has ever known, boyish even in old age. Hadamard says, I've been looking for you! Indeed he has made supper for them, a roast chicken so large that the table underneath has begun to sag. Hadamard smiles, not with his face but by engulfing the whole dream with his generous spirit. Sit down, sit down, he says, but there are no chairs.

The actual Hadamard would think for extended periods without words. His greatest difficulty, he said to one of his daughters, was translating his mental images into language, though over the course of a very long life he surmounted this difficulty time and time again, as is made clear by the trail of work he left behind. He published in the fields of:

> function theory
> calculus of variations
> number theory

analytical mechanics

algebra

geometry

probability theory

elasticity

hydrodynamics

partial differential equations

theory of gasses

topology

logic

as well as education, psychology, and the history of mathematics. He was a happy generalist, known for his good nature and what André called "an extraordinary freshness of mind and character." For twenty years, beginning in 1921, Hadamard convened a seminar on Tuesdays and Fridays at which visiting mathematicians presented their research, and often he would suggest an approach that a visitor had overlooked, or connect the topic to some far-flung province of mathematics not previously seen to be related.

But what did he think when he was thinking? What were those concepts floating free of words?

Hadamard was born in 1865 and died just shy of his ninety-eighth birthday, in 1963—a chronology I can barely wrap my head around, spanning so much of modernity—and he himself was quite interested in the question of what actually happens inside the brain of a mathematician. He was in his late seventies when he delivered a series of lectures in New York on the mental processes that underlie mathematical invention, which became the basis for his book *The Psychology of Invention in the Mathematical Field*.

Mathematical invention, he writes, is an instance of invention in general, one and the same engine underlying the creation of science, art, and literature. Intelligence is perpetual and constant invention, Hadamard said. Life is perpetual invention.

By the time the book appeared, all three of his sons were dead, two of them having perished in World War I and the third in World War II. He himself had fled Europe because he was Jewish. Only his two daughters survived him, one of whom would later recall that her father wrote his papers by dictating them to her mother, Louise, indicating with a "poum" that she should leave a blank space so that he could write in a formula later. He would say something like: "we integrate—*poum*—we see that the equation—*poum, poum, poum*—equals zero, takes the form—*poum, poum, poum, poum*."

And: Hadamard had a passion for mushrooms. Fungi and also ferns—once, he and Louise traveled across the Soviet Union by train, and at various stops the elderly mathematician would amble off to hunt for specimens outside the station, poor Louise fearing all the while that the train would leave without him.

Simone dreams of Crete. She travels to a village of ancient geniuses, who harvest grains and triangles, combine words and figures into a form of pure expressive speech, a giving back to the air in exchange for the gift of breath. They light fires on the beach. Fish swarm their feet in the shallows. Cherishing geometry and the ocean, they arrange their bodies in certain set configurations before they begin their open-air dances, which are cued by the rhythms of the waves. They draw pentagons in the sand, sing stories, wail for the dead, all forms of praise. The practice of an ecstatic order. They tell of the god who endowed them with speech as though he came by a few days ago. Yes, yes, they tell her, he was just here.

If only we had more access to the untranslated thoughts, to the mystery of how the mind churns. Hadamard, for his part, based his reflections partly on his own experience, albeit somewhat sheepishly. "I face an objection for which I apologize in advance," he wrote, "that is, the writer is obliged to speak too much about himself."

He drew from his wide reading and notably from a study undertaken by the editors of the French journal *L'Enseignement Mathématique*, which they published in two parts in 1904 and 1906. Called "An Inquiry into the Working Methods of Mathematicians," the study was based on responses to a long questionnaire they'd sent out to people in the field.

Sample question: *Would you say that your principal discoveries have been the result of deliberate endeavor in a definite direction, or have they arisen, so to speak, spontaneously in your mind?*

Another began: *It would be very helpful for the purpose of psychological investigation to know what internal or mental images, what kind of "internal world" mathematicians make use of . . .*

Hadamard asked similar questions of his contemporaries in the 1940s, and as an appendix to his book he reprints a letter from Albert Einstein. Like Hadamard, Einstein characterizes his thinking as something apart from language.

"The words or the language, as they are written or spoken, do not seem to play any role in my mechanism of thought," Einstein writes. "The psychical entities which seem to serve as elements in thought are certain signs and more or less clear images which can be 'voluntarily' reproduced and combined."

André dreams he is on a ship again, another ship like the ones that have lately carried him across the North Sea, across the Atlantic Ocean and up the eastern coast of the United States. He lies in bed yet his body remains in motion, rocking back and forth. Swelling and receding. But there is another bed in the cabin, and on it sits Daniel Bernoulli, the eighteenth-century mathematician restored to life. André gets up and follows him out onto the deck, gradually realizing that the boat is full of Bernoullis, the brothers, the sons, the nephews, also the unremembered sisters and wives of that sprawling, backstabbing mathematical dynasty. Bernoullis everywhere!

Were we privy to Hadamard's, or Einstein's, or anyone else's nonverbal thoughts, would we find that their subsequent translation into language served to crystallize those mental images into a more precise and elegant form? For a long time I thought of writing this way, as the art of making thought exact, of bringing latent insights and feelings into a fuller, more realized existence. Now I think that while this may be partly true, that some cloudy thoughts may be condensed into pools of precision, the longer I go on writing, the more I sense its limitations, see the tiny word critters scuttling around an inexpressible landscape.

Simone dreams her brother is a tooth—her own tooth, but not her own. Stuck inside her mouth and schooling her as always. She pushes at him with her tongue to wiggle him loose; although she doesn't want to be separated she still has that compulsion to dislodge him, to feel the bloody gap where he used to be.

Moreover, after many years of writing I find I have traveled down so many forking paths, motivated by a sensation that something like the truth awaits at the end of a path, only to wind up verbalizing the various ways that I have taken too many forks, been eluded by truth, realizations, epiphanies, et cetera, only to learn that the path either doesn't end or that it leads, as in a formal garden, into a cul-de-sac enclosed by hedges, where I come upon a moss-streaked stone pedestal that once supported a statue, the statue having been for some reason taken away.

———

Have you ever worked in your sleep or have you found in dreams the answers to problems? Or, when you waken in the morning, do solutions which you had vainly sought the night before, or even days before, or quite unexpected discoveries, present themselves ready-made to your mind?

André dreams of a bucket of candied fruit, one that he bought in actual waking life from a factory that was selling off its seconds, just before his family crossed the Atlantic. On the voyage they devoured those sticky scraps of pear and citron in the evenings, before all the passengers crowded into single-sex rooms to sleep in bunk beds, lumped together like litters of kittens. In the dream he puts his hand straight into the syrup, as others press around him, holding out plates or palms, touching his jacket, and though he would like to save all the sweets for Eveline and Alain, he can't turn these people down, no, in spite of himself he distributes shiny gobs of second-rate fruit to these mewling strangers.

Mathematical discoveries do not occur in dreams, Hadamard claims, or if they do, they are probably absurd. Yet Hadamard can't resist including a strange exception to the rule, reported by an American mathematician named Leonard Eugene Dickson, who had heard the story from his mother. When she was a girl, she and her sister had been keen on geometry, both competing against and collaborating with each other. They once "spent a long and futile evening over a certain problem," only to give up and go to bed, but during the night, in their shared bedroom, Dickson's mother dreamed of the problem and stated the solution, while asleep, in a loud and clear voice. Her sister heard her, got out of bed, and took notes. At school the next day, the sister gave this (correct) solution to the problem, which Dickson's mother, though she had dictated it in a dream, had no recollection of knowing.

In general, though, new ideas are far more likely to present themselves to a person who is just waking up, Hadamard notes, adding that he was once jolted out of sleep by a loud noise, and "a solution long searched for appeared to me at once without the slightest instant of reflection."

During the lull between waking and willing, the haphazard miracles of the liminal mind.

I can remember a dream I had in college—oddly enough, since I don't usually remember dreams the next day, much less years later—which was about matrices, rectangular arrays of numbers. The matrices of my dream were life-size, with detachable rows and columns that would hover over a person and act upon him or her in some inscrutable way.

In the introduction to an article published in 1990, the mathematician Robert Thomason explains that a dream ushered him toward the work he was presenting, and because of it he chose to include as his coauthor a friend and colleague who'd committed suicide the year before. "The first author must state that his coauthor and close friend, Tom Trobaugh, quite intelligent, singularly original, and inordinately generous, killed himself consequent to endogenous depression," wrote Thomason. But Trobaugh had returned to him in a mathematical dream: "Ninety-four days later, in my dream, Tom's simulacrum remarked, 'The direct limit characterization of perfect complexes shows that they extend, just as one extends a coherent sheaf.'"

Although Thomason, "awaking with a start," knew the idea was wrong, he pursued the argument, at dream-Tom's insistence, and discovered the path to the right idea in the shoals of the wrong one, landing quickly upon the results of the article.

Thomason himself died suddenly a few years later, of diabetic shock.

As far as method is concerned, do you make any distinction between invention and redacting?

The Latin *cogito*, meaning "to think," derives from a prior meaning that is "to shake together," notes Hadamard in a footnote to his book. Augustine had observed as much, he writes, and also that *intelligo* means "to select among."

Which is to say that cogitation is, at its verbal core, recombination and selection. Hadamard was a friend and admirer of the mathematician Henri Poincaré, and here he follows in the older man's footsteps. In a 1910 essay titled

"Mathematical Creation," Poincaré characterizes math as practically a form of spontaneous combustion, "the activity in which the human mind seems to take least from the outside world." As such, he says, it ought to tell us something about the essence of thinking.

What does mathematical creation consist of? asks Poincaré, who blazed his way through a large territory of mathematics and physics by relying on his remarkable geometric intuition. It requires not only the combining of existing facts but the avoiding of useless combinations: making the right choices. The facts worthy of study are those that reveal unsuspected relationships between other facts. Moreover, much of this combining and discarding and retrieving goes on without the mathematician's full awareness, occurring instead behind the scrim of consciousness.

Case in point: Poincaré's own struggle to prove the nonexistence of a certain kind of function. He recalls how every day he sat at his worktable for an hour or two, trying different things, with no luck. Then one evening he drank a cup of black coffee and couldn't sleep. "Ideas rose in crowds," he writes. "I felt them collide until pairs interlocked, so to speak, making a stable combination." By the next morning he had the outline of his results, establishing that a class of such functions did in fact exist, and he was able to promptly write up his work.

The prior banging of head against wall is necessary to the revelation, Poincaré insists: "These sudden inspirations . . . never happen except after some days of voluntary effort which has appeared absolutely fruitless and whence nothing good seems to have come, where the way taken seems totally astray."

The cruelty in all this is that the head-banging hardly guarantees the revelation, that to be an ambitious mathematician is to spend much if not most of one's time being stuck. Though maybe instead of saying *being stuck* I should instead say *chasing after tops*.

André Weil, in his description of the role of analogy in mathematics, those "slightly adulterous relationships" he conjured in one of his letters to Simone, may have valued the chase over the capture, writing that "the pleasure comes from the illusion and the far from clear meaning; once the illusion is dissipated, and knowledge obtained, one becomes indifferent at the same time."

The flicker of a parallel, the suspicion of a connection, excited him, more so than nailing it down, working out the details. As though knowledge itself were a bit of a letdown: it's being on the cusp that brings the greater thrill.

André dreams himself back to the École Normale Supérieure. What ease he feels in Paris, home at last, making his way along the riverbank, then down an alleyway and finally to a long, narrow set of stairs, which zigzags up the side of a building made of large white bricks. There, on a high balcony, friends he doesn't recognize are waiting for him because he is supposed to lead a seminar on the topic of "Bethlehem fields." When he wakes up he still has the term in his head, reaches for the definition before he remembers that there is no such concept and that he is living in an American town called Bethlehem, Pennsylvania.

Simone dreams she has developed a new physical theory, of continuous atomic states, freedom of the electrons to assume any energy level whatsoever. Gradually she becomes aware that she hasn't simply invented a theory but has imposed it on her surroundings, which turn more and more diaphanous, which begin to dissolve before her dissolving eyes.

Would you say that a mathematician's work should be interrupted by other occupations or by physical exercises which are suited to the individual's age and strength?

One night after having watched an algebra lecture, I dream that I am standing in front of a blackboard, next to a woman I've played pickup basketball with (that is, a woman I identify as the basketball player, even though my dream version doesn't look much like her, not least because her dream hair is bright orange), and I point to a symbol on the board and remind her that the subgroup under discussion is a normal subgroup because it's the kernel of a homomorphism.

Poincaré speculated that the elements of thought were something like the hooked atoms imagined by the philosopher Epicurus. When the mind is at rest, our thought atoms remain as if hooked to a wall, stationary, and so never meet, but by thinking we agitate them, *cogito*. As a result they collide, interlock, and surprise us.

Several of his discoveries leaped to mind while he was walking someplace, or, in one case, while stepping onto an omnibus.

Said Georg Cantor of one of his own results: "I see it, but I do not believe it."

Another night I dream that I'm back at college, only it's not 1990s college but rather Radcliffe College circa the 1950s, a distinctly black-and-white world as in an old movie, and I'm at a tea where the special guest is André Weil. He's there to speak about his work, but at the moment he's drawing an analogy between something in math and a woman's breasts, he's smirking and positioning his hands in front of his ribs like he's supporting the weight of some serious mammaries. The few men in the audience chortle, while the others, the women, are unimpressed. I can tell I'm the only one who wants to cut André some slack—that is, I want the other women to see past the off-putting delivery and appreciate the math.

There's a question-and-answer period following the talk, and I think of a question that I believe will rescue him, a means of illuminating something about his work, but I'm not called upon; instead the woman sitting next to me is, and in response to her question Weil leans toward her and tells her she has beautiful eyes. After that, as people begin to cluster around him, he starts for the door, pushing through the crowd, and like a reporter trailing an accused man away from the courthouse I catch up to him and yell out a question.

It's not the one I came up with earlier, though. "How do you pronounce your name?" I shout. Confused, he mutters something, and I say, "Your name, your last name!"

"Vay," he says dismissively, and marches off.

Naturally the mode of mathematical thinking varies by thinker. According to Hadamard's informal survey of colleagues in the United States, George Birkhoff would visualize algebraic symbols. Norbert Wiener would think either with or without words. For George Pólya, one word might appear in his head, and ideas would precipitate around it. Hadamard goes so far as to specify the "strange and cloudy imagery" that arises in his own mind as he follows a simple proof about prime numbers, listing the steps of the proof on the left-hand side of the page, his mental images on the right. The images are, for example, "a confused mass," or "a point rather remote from the confused mass." In fact, he writes, every time he undertakes mathematical research he develops a set of such images, which helps hold everything together.

Poincaré believed he'd inherited his mathematical talent from his grandmother.

———————

Do you experience definite periods of inspiration and enthusiasm succeeded by periods of depression and incapacity for work?

For much of his life Cantor suffered from depression and was repeatedly hospitalized. In the late nineteenth century he came up with revolutionary ideas about the concept of infinity, which his contemporaries viciously cut down. Beginning in middle age he also dedicated himself to trying to prove that the true author of Shakespeare's works was Francis Bacon.

Simone dreams she is in a refugee camp, relieved and excited to have reached a place where she might at last suffer boundlessly. She throws away her shoes, then goes looking for a friend she's never met. What does he look like? How will she know him? By his purity, she thinks. One half of the camp is forest, the other desert. She heads straight for the desert in bare feet that are already starting to burn. Finally! A world cleansed of luxury and split open by such fierce light.

André dreams he can no longer dream. A doctor—his father?—a doctor who doesn't look like his father but is nonetheless, André understands, Dr. Bernard Weil, gives him the diagnosis. It is just the two of them in a small, square box of a room with metal walls, a metal ceiling. But, but: the objection hovers just out of his reach, he almost knows and is almost capable of presenting the obvious counterexample to this theory that he can't dream, but he doesn't quite have possession of it, and rage rises up in him, he wants to strangle the doctor who is and is not his father.

If any persons who have been well acquainted with defunct mathematicians are able to furnish answers to any of the preceding questions, we ask them instantly to be kind enough to do so. In this way they will make an important contribution to the history and development of mathematical science.

"I am trying to answer in brief your questions as well as I am able," Albert Einstein wrote to Hadamard. "I am not satisfied myself with those answers and I am willing to answer more questions if you believe this could be of any advantage for the very interesting and difficult work you have undertaken."

8. "MY DEAR BROTHER," SIMONE IN France writes to André in America. "My attitude has not changed. I don't wish to live in America for a whole pile of reasons." She and their parents are sharing an apartment in Marseilles, and while Bernard and Selma intend to emigrate as soon as possible, Simone wants to remain in France, to suffer whatever the French are made to suffer—"even if a real famine occurred, I would undergo it like the others . . ." André's case, on the other hand, is different; she thinks he's where he ought to be, for the sake of his work. "A mathematician is so rare an animal that he deserves to be preserved, be it only on the score of curiosity."

She would go to America only on one condition, and that's if she could be sure that it would help her realize her ambition to organize a cadre of front-line nurses. It's a fixation of hers, a proposal that nurses be dropped by parachute onto the battlefield to treat the injured and dying, though many would surely become casualties themselves.

A flock of kamikaze benevolents, floating above the front, landing among the bleeding, the legless, men drawing their last breaths. They would bandage the wounded and offer comfort, at least until they themselves had their heads blown off. Simone would be among them, naturally. She is desperate to see the plan carried out, petitions officials, brings it up over and over, her fantasy of sacrificing herself converted into a patriotic mission—though not so much patriotic, in her mind, as humane. To plant nurses right in the middle of the slaughter would be an act of defiance, a strike against inhumanity itself.

Reading Tacitus, she learns that it was customary among ancient Germanic peoples, the barbarians and seminomadic tribes of the first century, to bring a young girl, surrounded by elite warriors, to the front line of battle.

It's hard to see how Simone could've been so serious about such an untenable idea. All I can hazard is that she had hit upon a notion so perfectly in tune with her psychology, one that so encapsulated her search for meaning in affliction— her desire to transmute her own alienation into difficult charity and at the same time to summon her own annihilation— that she refused to consider its flaws.

As time goes by she gravitates more and more toward Catholicism, though it wouldn't be quite right to call her a convert, since (as she explains to the priests whom she ropes into scholastic debates) she's unsure whether she could truly belong to the church, whether she even wants to belong. Her stated preference is to suffer alongside the sinners in the wilderness, though in practice she's more like someone standing just outside the church door, hectoring the priest with questions.

In Marseilles, the terrible housing conditions of Indochinese laborers rile her; she also agitates for refugees from neighboring countries and corresponds with a Spanish anarchist interned in a French camp. More often than not, it seems, the people who inspire her moral outrage are a step removed, different from her. Publicly, at least, she doesn't express the same kind of alarm about, say, injustices endured by women. She hardly mentions the plight of Jewish people; later she'll write that the best course for Jews would be to intermarry and assimilate.

In high school I indulged in a more conventional dream of an airborne vocation, thinking I might become a foreign correspondent. This idea may have been shaped in part by Margot Kidder as Lois Lane, climbing up the elevator shaft of the Eiffel Tower in *Superman II*, and even more so by a certain credit card commercial that used to play on television: a man and a woman living in different places fly to meet each other in some European capital for the weekend (a trip made possible by the credit card, I guess), and although I don't know whether the pretty actress in the ad was supposed to represent a newspaper reporter, that's what I thought she was, because, as I remember it, she was wearing a trench coat.

So unusual in Pennsylvania is the aurora borealis that when the sky becomes streaked one summer night with plumes of light, like smoke from some extraterrestrial volcano, and the horizon below turns bright pink, people suspect that the

Germans are behind it. André knows better, he's seen a show of lights like this before, in Finland, but it seems impossible that they would be visible so far south, and so he is invaded by the sense, however absurd, that what they're witnessing in the sky must be a portent. Portent of what? He sits in silence with Eveline and Alain on the porch of their rented house, all three of them dumbfounded by the spectacle. He pictures his parents, his sister, asleep in Marseilles; the separation, the guilty realization that it's probably been a day or two since he thought about them; the smallness of his family, the shrinking of possible outcomes and at the same time the tiny being now growing in his wife's belly, the child of them all.

After I went to college my dreamy notion of becoming an expatriate journalist seeped underground, but from time to time some version of it would bubble up, even as I skirted around the math department. Most of my math courses followed a standard format—lectures pegged to a given textbook, problem sets, exams—but I took one class, a seminar on dynamical systems, that attempted to combine the mathematics and the history of the subject. We were assigned to write a paper, in addition to solving problems. One day I found myself in the professor's office. I don't remember whether I'd gone there on purpose or just happened by the office while he was sitting there with the door open, but I think it was the latter: I walked past and he called me in to talk. He complimented the paper I'd written and asked whether I was planning to apply to graduate school in math.

I assumed he was encouraging me because the math profession could use more women, but what he said next was that the math profession could use more writers. "We need people who can explain this stuff," he said.

I gave a noncommittal answer, I wasn't sure, maybe, but what I remember thinking was: I want to be a *real* writer. I'm not going to write about *math*.

Very early one morning, policemen knock on the door of the Marseilles apartment. Selma Weil answers, then goes to wake up her daughter, who rouses herself and dresses quickly. As Simone sips the coffee her mother has made, the police ask about the local Resistance network, which she recently joined in hopes of being sent to England. She tells the men she doesn't really know the others in the network, doesn't recognize the people in the photos they show her, doesn't remember names. She says she learned about the group from a person at a bus stop whom she never saw again. The police search her room, which is so full of manuscripts that Selma finds herself feeling sorry for them.

On three occasions Simone is summoned to the police station for questioning. Each time, in the belief that she'll be arrested, she brings a valise packed with clothes and a copy of the *Iliad*; each time her parents accompany her and wait across the street at a small café while she is inside the station. Her interrogators leave her in a corner and speak to her only occasionally, issuing vague threats of what might happen if she doesn't inform on other members of the Resistance. She stalls and stalls.

At last an asthmatic officer tells her she's a little bitch.

You know I could have you thrown into jail with a bunch of whores, he says.

She can see her parents out the window, anguishing over plates of untouched food.

I've always wanted to know that environment, she replies. Which is the absolute truth. She returns the policeman's stare, and in the end she is let go.

What was I thinking in that professor's office, what did I mean by "a real writer"? I remember a conversation I had with a fellow student in the math summer program I attended after freshman year, a computer and math nut who always wore tracksuits and who'd grown up on an island off the coast of Washington State, who'd only ever been to the movies on group excursions during this summer program, an American singularly removed from American culture. We were eating lunch or maybe dinner in a university cafeteria, and I found myself arguing with him, pestering him to admit that he hoped to write a novel eventually, which I believed everyone secretly did. He explained that, to the contrary, his intention was to pursue a Ph.D. in math.

"But after that," I was saying, "surely one day you want to write a novel . . ."

"No!" he said, exasperated. "I want to be a mathematician!"

Finally, reluctantly, I was persuaded that he had no interest whatsoever in writing a novel. Only after this conversation did I begin to think that my desire to be a writer was something particular to me, that even if many other people also wanted to be writers, and even if there was a part of me that recoiled at the whole soggy notion of being an aspiring writer, nevertheless I was, already, an aspiring writer.

According to one scholar's thesis, it was the invention of writing that gave rise to number as an abstract concept. The prehistoric people of the ancient Near East, exchanging sheep or grain, originally recorded what they'd traded using clay tokens that represented the thing traded; in time they began storing the tokens in a type of envelope, marking the envelope to designate what was inside of it. Eventually they dispensed with the tokens, in favor of the marks.

Simone has arranged to meet, at a café in Avignon, a writer-farmer: that is to say, a self-taught philosopher who owns twenty acres of land with a stone farmhouse where he lives with his wife and his father. This man, Gustave Thibon, is a devout Catholic and a meticulous diarist. Sitting near the café's entrance, he waits for her and notices the rot at the bottom of the doorframe that she has yet to walk through, the wobble of the table. He's uneasy about the whole thing, about this woman who wants to be hired to do manual labor. Yet his friend Father Joseph-Marie Perrin asked him to consider bringing her on, and in deference to the priest he has agreed to host her for a few weeks.

A philosophy teacher from Paris, the priest told him, living in Marseilles for now. An intellectual, a woman of faith.

Thibon said he couldn't understand why she would want to work on a farm.

Truth be told, I don't entirely understand it myself, the priest said.

When she enters the café it seems to him as though a moderately strong wind could spirit her off. A shipwreck of beauty, he will later recall her as, careworn and concave. She's barely over thirty but looks much older and at the same time ageless. In spite of the summer heat she has on baggy wool pants and a jacket over her shoulders. An untended mass of black curls. Her eyes, scanning the room and then finding his, startle him, and right away he thinks: She's possessed. Her very presence causes him a peculiar kind of pain.

She disagrees with him about everything, countering whatever he says in an oddly monotonous voice, until he's practically drained of speech. Two hours they spend hashing out the terms of her employment, which she fears are overly generous.

I don't care to have any privileges the other workers don't enjoy, she insists.

All I meant to say is that they have more experience, and if you become exhausted you must—

I will keep on. I'll take my breaks when the others do, or less often if necessary.

If necessary?

To do my share.

He avoids putting her to work until she insists he stop treating her as a friend. In the fields she's an even less promising laborer than he anticipated, blundering and overeager. More than that, she's a source of worry. How much can she take? What if she collapses? Yet she has a doggedness, a will to keep toiling—and there's something about that pure gaze of hers that makes him respect her.

In the evenings, they read Plato in Greek together.

She refuses to accept a bedroom in the farmer's house, protesting that it's too comfortable for her. Instead she settles into a tumbledown cottage at the edge of the woods, its dirt floor scattered with old rat droppings.

(All it lacks is a door of tar.)

She calls it her fairy-tale house, and so becomes the sort of figure—a sprite, a witch, a shapeshifter—who would live in one. She fetches her water from a nearby spring, sweeps out the droppings, clears weeds outside, gathers wood, makes fires. She eats very little, wishing she could be like one of those hibernating animals who require no food. She sleeps on the dirt floor.

"I continually see the light of the sun shine in a different way on the valley and the hills," she writes in a letter, "and then, at night, vast stretches of starry sky."

Her parents come to visit and encounter an old woman who, not knowing their identities, exclaims that the man who owns the vineyard has taken a mistress: "She is a crazy woman who lives in a hut!"

There is something tempting, albeit ultimately unsatisfying, about the theories of the semioticians, those who would leave reality out of it and explain math as simply a language, a sign system, an intersubjective method of persuasion. If only for selfish reasons—then I could conceive of what I did in college not as a random detour but as a specialized kind of writing, a strange and unnecessary but not entirely irrelevant preamble to all the other writing.

André has taken an underpaid instructorship at a college in Pennsylvania, a position he considers beneath him and would not have accepted but for the fact that he has a family to support, a baby on the way, and no better option. (Although a mathematician of rare talent, he is also a Jewish foreigner—it hasn't been easy to secure an American job.) The college is a second-rate engineering school that trains young men to go to work for Bethlehem Steel. A diploma machine: he becomes a cog. His colleagues know nothing of genuine mathematics; their task (and his) is to feed formulas to narrow-minded students, young men he considers mediocre, dismally incurious—he isn't exactly sensitive to the fact that none of them grew up with his privileges, they are working-class kids trying to make their way up the ladder, now with the draft hanging over them. He doesn't foresee that some of them will be sent to his own country to fight in the war that he escaped.

Once in a while he forgets where he is, forgets the dull colleagues, the rows of blank faces in the classroom, the clicking of the mechanical pencils, the steel mills and their black smoke. He launches into a proof, an argument he finds beautiful and, flicking his wrist like an orchestra conductor, he marks out lines of symbols on the chalkboard, losing himself in the chain of reasoning.

This completes the proof, he concludes, still holding his

stub of chalk up in the air, still in his bubble of contentment as he turns back to the students.

A vast silence follows.

Are there any questions? he asks.

Always the same one. Is that going to be on the exam?

In an interview posted to YouTube, the mathematician Maria Chudnovsky discusses how she and three colleagues developed a 150-page proof of something called the strong perfect graph theorem: "You know, when you work on a problem like that where the solution is long, first you just try in the dark, you try different things. And then you start to get an idea of how things should go. And then sometimes you think you've got an idea, but in fact you're wrong. But then there comes a time when you imagine some sequence of steps in your mind, and you're able to take all the steps."

You try in the dark, you develop an idea that's maybe the wrong idea, indeed you might run down a bunch of wrong ideas, but then (if you're lucky) another, better idea comes and you're able to imagine your way to a solution. That is the fundamental narrative of the creative process in any field. And of writing in particular, because (like math) it is so interior, theoretical, on some level a game with symbols. However much these games depend on an external reality, receiving inspiration from and following rules imposed by the world outside our heads, they require barely any equipment, they can be played by a person alone at her desk.

At her request, Thibon helps Simone get hired at another farm for the grape harvest. There she will kneel and cut grapes with an old pair of steel clippers, laboring for hours, past what should be her breaking point, eventually so depleted that she lies down under the vine trunk but still reaches up to tug another bunch toward her face. (Sometimes I am crushed by fatigue, she admits to a friend, but I find in it a kind of purification. She writes to her parents that the work brings her great joy.) When she goes to sleep at night, it's as though she is still at work, cupping the fruit and snipping the stems. Her body becomes sticky and stained from grapes that have burst, and her fingers won't fully extend, as though they've been snared in a gummy purple web, but she must go on with her cutting, go on inching down the rows with her back in the dirt, her hair in the dirt, dragging her wooden basket along, the marvel of everything made—the basket, the clippers, her shoes, her hands—not lost on her even as she is picking grapes in her sleep.

One day, as she'll tell the farmer, she wonders whether she hasn't died and gone to hell without noticing, hell having turned out to be an eternal grape harvest.

She memorizes the Greek text of the Lord's Prayer, recites it to herself until the words tear her thoughts from her body and deliver them to a place that seems to be outside of perspective or point of view, outside of space itself.

People who encounter her during this time, old friends and strangers alike, are struck by her presence. A nun, one Sister Colombe, would recall meeting her—how she entered a room wrapped in a navy-blue cape and stood quietly behind a companion—and would remember the indelible impression made by her manner of silence, the quality of attention that radiated from her at first sight.

For a spell she becomes unhitched from any regimen or routine. She writes, "I have fallen into a kind of abyss in which I've lost the idea of time."

It's probably best to leave analogies between math and writing in the semi-obscure territory that André Weil described in his letter to his sister—that realm of intuition, furtive caresses, mental courtship. But I think back to the conversation I had with the young mathematician who I believed must want to write a novel, and while it's funny to me now, the way I projected my own secret hope onto someone sitting across from me at lunch, I think another reason I was convinced that a math person would wish to write fiction is because he was a math person, already inclined to probe and to help extend a kind of alternate universe. I wonder whether mathematicians and fiction writers might be people for whom the lure of alternate worlds is particularly strong. And then I wonder whether this is just a natural consequence of abstract representation itself, that once you start putting words or numbers on paper you are already beginning to piece together a kind of parallel universe, which you then want to access, discover, flesh out. I might go so far as to conjecture (too broadly, too tentatively) that even hash marks on ancient clay envelopes contain the seeds of this desire.

André, across the Atlantic, has become more tightly roped to the idea of time, as the expansion of Eveline's stomach reminds him. His wife has become so large he's sure there must be a giant baby in there, or three babies, no other pregnant woman has ever been this huge, he's convinced, though in reality he's just never paid much attention to pregnant women, much less shared a bed with one.

Once the baby is born she seems to him as little as Eveline was big. Sylvie, they call her, not actually intending to recombine pieces of her mother's and her aunt's names, yet there will be times in her life when André starts to address her as Simone before he catches himself.

While mother and baby are still in the hospital, his friend Chevalley, one of the Bourbaki crowd who's now teaching at Princeton, drives over to help out at the house and talk mathematics. As André washes dishes Chevalley stands at the drying rack, listening to André discuss the problem of counting solutions to polynomials, which he punctuates by waving a soapy glass in the air, letting off a flurry of suds. His stepson, now twelve, watches as André hands that same glass to Chevalley without having rinsed it off, then plunges his sleeve into the dishwater, not noticing until after he's finished saying what he's been trying to say about Dedekind's generalization of the Riemann zeta function. Quite a few plates and cups and pots, washed after previous meals, are sitting on the countertops, as André has not bothered to put them away, in some cases doesn't even remember where they go.

When he brings Eveline and their daughter home, the baby seems like something else he might all too easily lose track of, her forehead still wrinkled, her hair a matted black skullcap. Her arms flail and her tiny fingers curl. Her doll's lips are wet. When he picks her up she is all but weightless, how can it be, a living human.

One evening, a few days before Easter, Simone travels by train to the town of Carcassonne to visit a writer named Joë Bousquet, who during the First World War took a bullet to the spine that completely paralyzed him. For twenty-two years he's been confined to bed; he continues to live with great physical pain.

The night is lit by a low yellow moon. Simone reaches the house very late and pushes gently on a door that Bousquet's mother left open, finding a clean, spare house with almost no furniture, only two chairs and a table and a lamp in front, and beds in the bedrooms, one of which belongs to Bousquet.

He is awake. He never sleeps for more than a few hours at a time, he whispers. He'll need to be moved but sometimes there's no one to move him, and then all he can do is meditate on the crushing pains in his legs and in his sides and in his back where the bullet shattered his vertebrae.

Simone is on her knees. She asks him to tell her more, to describe his pains to her. He meets her eyes and calmly obliges. Sometimes I am shot again, he says, I feel the bullet rip into my spine again. Sharp glass splinters explode all through my body, I am cut everywhere. Sometimes my joints feel like they're on fire, a fire deep inside my legs. Or nails are driven into my hip bones. Every day and every night is different, he says.

It must be torture, she says, as much in fascination as in sympathy.

It is, but I've learned how to observe pain. I see it happening in myself.

What is it like, to watch?

It has . . . I can't put it into words. I can only watch.

I wish I could know what you feel. Maybe it sounds naïve.

Yes. But I think I understand you.

They go on talking for hours, until three or four in the morning. Refusing a bed, she retires to the next room to sleep on a mat.

Later she'll write a terrifying prayer in a notebook:

Father, in the name of Christ grant me this. That I may be unable to will any bodily movement, or any attempt at movement, like a total paralytic. That I may be incapable of receiving any sensation, like someone who is completely blind, deaf, and deprived of the other three senses. That I may be unable to make the slightest connection between two thoughts, even the simplest . . . That I might be insensitive to every kind of grief or joy, and incapable of any love for another being, another thing, even for myself, like old people in the last stage of decrepitude.

At long last Simone and her parents receive permission to travel to Casablanca, where they spend seventeen days in a camp for transients. The sun beats down; there aren't enough chairs; refugees fidget and quarrel and piss into the sand; they snatch up the newspapers, any morsel of information like scarce food. Simone, suddenly territorial, occupies a chair all day long, and when she must get up, one of her parents will sit in it, but only to save it for her. She writes all day, hard at work on an essay about the Pythagoreans.

From there they board a ship bound for Bermuda, then New York. The crossing lasts a month, and the other cabin-class passengers, released from the deprivations and anxieties of war, indulge themselves in every way. Simone, irritated by all the carousing, tries to sleep in steerage, then on the deck. She calls the boat "the floating brothel."

André meets the ship when it docks in New York. The wharf is raucous with the blasting of ships' horns, and bleating gulls, and paperboys, and he fears he won't find them, he searches among the descending passengers for his family until he sees his parents, two little gray birds stepping gingerly down the gangplank, threading their way through masses of larger people. He doesn't see Simone, until abruptly she comes tearing toward him.

Her skirt is rumpled and her eyes shine out from hollows of skin and vein. They've barely said hello before she starts in on the possibility that she might be able to meet with President Roosevelt to discuss her plan for front-line nurses, and she talks so incessantly he finds himself losing track of what she's saying. In another family, less accustomed to obsession, Simone might be considered insane. Another family would say that she'd lost her mind.

PART FOUR

9. OVER THE COURSE OF THE 1940s, beginning with his time in the Bonne Nouvelle prison and then in the years that follow, during which he emigrates to the United States and becomes a father to two daughters and loses his sister, André develops a set of ideas—proposals supported by solid theorizing but not yet proved—that become known as the Weil conjectures. They develop the analogy that he described to Simone in his letter from prison, relating the fields of topology and number theory. Since topology is a mathematics of space and number theory concerns the properties of whole numbers, it's far from clear that one should inform the other. But the Weil conjectures establish a surprising link between them, and so span the gap between the continuous (space) and the discrete (whole numbers), seemingly quite distinct and yet somehow part and parcel of the same truth.

These conjectures are outside the limited terrain I covered in college. How little, really, I learned in college, less and less every semester, since whenever I studied one thing, all sorts of further potential studies would reveal themselves, so that instead of making progress I only walked a short way up a sand dune that continually grew larger as I went along.

No sooner has she made it across the ocean than Simone wants to go back home. Her hope in coming to America was that she might somehow find support for her airborne nurses, but that dream doesn't last long. I can't go on living like this, she tells her parents.

She spends hours in the waiting room at the French consulate, intent on convincing an official to send her back to Europe, to work for the Free French. A fixture in yet another waiting room, silently chain-smoking, scribbling in a notebook.

She takes a nursing course in Harlem.

In grade school we are introduced to the number line, presented as a black horizontal strip. Arrows at both ends tell us that the line extends infinitely to the left and infinitely to the right, the positive and negative numbers running off in opposite directions, the zero presiding coolly in the middle. Over the course of however many years of school we populate this line, first with whole numbers, then with fractions, then with irrationals, like $\sqrt{2}$ and π, and arrive at what are called the real numbers.

A strange, bold term, real numbers, as though we could hold π in our hands. Sometimes they're called simply "the reals."

Now (bear with me) consider just a section of that line, for example, the interval [0, 1], stretching from 0 to 1, containing 0, 1, and all the infinitely many numbers in between.

And let's say you have a map $f(x)$, a function that takes any value of x in [0, 1] to another point $f(x)$ within the interval [0, 1]. Starting from that section, the function will land us back in the same section.

There's a theorem, one that's not hard to demonstrate in this one-dimensional case but that I still find pleasing, about any such function that is continuous, meaning roughly that if two points x_1 and x_2 are sufficiently close together, then $f(x_1)$ and $f(x_2)$ will also be relatively close together. In that case, there will always be at least one point fixed by this map. There's always at least one x for which $f(x) = x$. No matter how you act upon the interval, no matter how you twist and torture it, there will be one or more points left in place. And there are analogues to this in two or three or any number of dimensions.

An extension of this result, in all finite dimensions, was the Brouwer fixed-point theorem, and after that came the Lefschetz fixed-point theorem and a theory of how (roughly speaking) to count the fixed points of continuous maps. Unlike the great majority of math theorems, which I take to be strictly about math—that is to say, I don't think of them as suggestive of anything beyond math—in my mind the fixed-point theorem ripples outward. Even the most abstract of transformations, the mathematician's perfect, disembodied maps, cannot change everything at once. Although I'm now resorting to loose, messy analogy, I'm abandoning rigor in a way that might make a proper mathematician cringe, this inclines me to think that any transformation in any realm has a fixed point; I think of Kafka's Gregor Samsa—of the part of Samsa that remained Samsa after he became a cockroach—or of the pieces of myself that remain constant over time, in spite of so many physical and personal alterations, in spite of the fact that over the course of however many years every cell in the body renews itself, and while none of this is in fact even remotely implied by the Brouwer fixed-point theorem or the Lefschetz fixed-point theorem or any other theorem, I nevertheless associate the idea that not everything changes at once, that in the middle of no matter what catastrophe there's something that stays still, with the theorem's $f(x) = x$.

And then again I am tempted to say, even more loosely and unforgivably, that math itself seems to me a fixed x, a still point surrounded by human frenzy, the eye of the storm.

Although this hardly serves to justify my gauzy appropriation of the Brouwer fixed-point theorem, I'd like to note that Luitzen Egbertus Jan Brouwer, a Dutch mathematician born in 1881, was himself drawn to mystical thinking. He had to be persuaded by his adviser to cut a section of philosophical musings from his dissertation, and on top of the math lectures he delivered to students in Amsterdam, he gave a series of more-wide-ranging talks later published in a book called *Life, Art, and Mysticism*.

He married a divorced older woman, and they built a weekend cottage in Blaricum, a proto-hippie town by the ocean, where they could pursue their interest in herbal medicine and practice nudism. As he grew older Brouwer invited many fellow mathematicians to Blaricum, and though maybe he wore clothes during those visits I still imagine otherwise, the renowned topologist serving his guests a bitter tea made from roots or sticks and offering his thoughts about homotopy theory, naked all the while.

When Simone comes to visit, Eveline expects that she might be tentative with the baby, but no, not at all. "How beautiful you are in your angel's robe!" she crows as she shifts Sylvie into her hands. For a moment Eveline and Simone both hold her, Eveline not quite sure, not quite ready to let go, and when she does release the baby she has to check the impulse to instantly grab her back because of the way Simone has taken possession. As though staking a claim. Simone draws her niece's soft lump of a body to her bony one and nuzzles the baby's head, her dark curls spilling down over tiny Sylvie.

That a person could ever be so light and vulnerable, it's impossible to reconcile with the weight we suppose ourselves to have as adults. It exposes the illusion of that weight. Now Simone's lips are moving, muttering something to the baby, more intentional than simple coos and hellos and yeses. Verses, maybe, or prayers?

Sylvie starts to cry, and Simone lifts her head, not all the way but just enough to reveal a curious, strained expression that Eveline has never seen before. She's nearly cross-eyed. It's as though rather than comfort Sylvie, she's trying to absorb the crying into her own body. The baby pauses, they all pause, four hearts beating. Then Sylvie starts to cry again.

Eveline thrusts out her arms. How reluctantly Simone surrenders. She has never looked worse, Eveline thinks, and she has never looked more alive, the way her eyes radiate out of her caved-in face.

Simone and her brother trade short phrases in what Eveline assumes is ancient Greek. Half the time she can't even follow what the two of them are saying. It comes as a relief that there is another person in the room, alien creature though this baby may be, an ally nonetheless. Even as her tiredness and the bottles and the diaper changes make it all the more impossible to keep up with the talk between brother and sister, now it doesn't matter—so be it—for here's this girl who is her own, hers and André's.

Isn't she? Only look at that mat of ebony hair on Sylvie's head, soon to become curls, exactly like her aunt's.

The seeming fixedness of mathematics is surely one of the reasons I've felt drawn back to it, given our present-day world's particular instabilities and alternative facts, but another reason, a stronger reason, is that my son likes math, which is not to say that I need to relearn abstract algebra for his sake but rather that his excitement has reminded me of my own old excitement, has made me want to blow on the embers—has made me realize there are embers. And as I do, what strikes me are the dialogues, the exchanges, whether it's me talking with my son about numbers or Benedict Gross's performance of algebra. Even as mathematics presents itself from afar as an austere architecture dreamed up by singular geniuses, up close it's a torrent of transmissions, teachers lecturing, college kids trying to solve problems together, colleagues at conferences, André writing to his sister. For every solitary discovery there are massive systems of relationships, which I begin to think of as a kind of giant math ant colony, or math hive, and I even begin to wonder whether (or conjecture that) the desire for mathematical revelation, the wish to dwell in a perfect, abstract world, is secretly, unconsciously twinned by another desire for communion. One the negative imprint of the other. Abstraction the flip side of love.

"Nothing which exists is absolutely worthy of love," wrote Simone.

"We must therefore love that which does not exist."

For three days she stays with him, and for three days they argue. Few people listen to her like he does. He has always felt it would be cruel to tune her out, and for that matter he likes the challenge, stress-testing her intricate arguments, objecting to them, developing them further. Even when she crawls far out onto a rhetorical ledge that's beginning to crumble underneath her, still he goes along, keeps her company until she's ready to come back.

His armchair engulfs her as she glares at the ceiling, smoking, still wearing her rumpled black coat; its wide collar, flecked with ash, droops down from her shoulders. I'm through with this, she says, I should've never left France.

You're safe here.

If I'm forced to stay I will go south and work in the fields with the blacks. I don't care if I die there.

I think the fieldwork here is more difficult than in your Catholic vineyard.

Good, I hope so.

You might harvest tobacco, he muses, as more ash drifts onto her coat.

Yes, you're right!

It's settled then, he says, and sighs.

She ignores his irony and nods once. May I give Sylvie her bottle? she asks.

Simone can't get enough of feeding the baby; she loves to watch the spot beneath Sylvie's chin that pulses in and out as she swallows, loves to hear the little clicking gulps. Meanwhile Simone eats almost nothing herself, nudges her food with the very tip of her fork.

Isn't there some way to serve the war from here? asks André.

Doing what? Knitting socks for soldiers? You know I'm hopeless for that sort of thing.

A train lumbers by, laden with the steel beams that are forged in Bethlehem and sent off into the depths of this country that is continually building itself.

There seems to be plenty of factory work, he says, only half kidding.

I'm happy for you and your family to know this kind of peace, but I have a different purpose.

What is it?

If my plan for the nurses is not approved, I am going to request that I be sent into France on a mission of sabotage. Even if it's one I won't survive.

Half in love with easeful death, he says in English.

It's not only you who know your own dharma.

He's quiet.

For almost the entire hour before she leaves for the train station, Simone stands over the bassinet, gazing at Sylvie as she sleeps, whispering to her, bringing a hand to her chest to feel her breathe.

It will be another six weeks before Simone's British visa comes through. On the phone with André, or in the apartment with her parents, she'll talk about the baby incessantly, her normally slow, severe voice becoming faster and higher pitched. The lovely Sylvie, the delectable Sylvie! She's nearly as besotted as Eveline and André, or maybe more besotted, since she can love the baby purely, a distant crush, without being exhausted by her. She'll warn her brother—though Sylvie is still an infant!—not to let her become a flirt.

Just as she is leaving her brother's house, stepping out the door with her bag full of books, Eveline says to her, We'll be seeing you, Simone.

Behind her glasses Simone's eyes narrow. A sad pitiful smile spreads across her face. No, you won't, she says. This is goodbye.

10.

BETWEEN NOT KNOWING AND KNOW-
ing, what is there? A doorway, a slim
threshold—or maybe a dark, rocky path
joining distant points: the journey from one end to the other
might take years, it might take centuries.

A conjecture forges a trail, shines a torch and clears the
initial stretch. It's a setting forth, full of plucky confidence,
and at the same time it's a reminder that the destination is
not yet in sight and might well be unreachable—an aspira-
tion that may never be realized, an arrow spinning in the
wind.

"I have had my results for a long time, but I do not yet
know how to arrive at them," wrote Gauss.

I take it back: the region between the unknown and the
known isn't really a path, better to call it a wide expanse with
very few directional markers, a field through which you beat
your own track, uncertain of where it will lead and barely
noticing how that choice of one track causes other tracks to
disappear. All but oblivious to the quiet but tremendous collapse of countless other possible routes, avenues that will go
untraveled as other theories are not developed, as all the
possible books I might have written, perfect in their nonexis-

tence, are replaced by this flawed one. (Or take this sentence itself, knocking off whatever else I might have come up with, as I keep feeling my way toward an unknown that will only ever recede.) Along with the optimism of conjecture, that faith in what seems true, perhaps there is or there should be a distinct unease, a regret for all that's been lost in the process—or maybe one of those compunctions that come upon me in dreams, that sense of having forgotten to do something. In this case having forgotten to regret what was lost in the process.

I'm leaving next week, Simone in New York writes to André in Pennsylvania. Impossible to go to Bethlehem. I'm terribly busy. I so regret not being able to say goodbye to Sylvie. I hope to see her again before she's married. In any case I've cast a spell on her, you'll see in a few years.

André reads the letter in his living room, standing by the window with baby Sylvie in the crook of his arm. When he's done reading he shifts her into his two hands and holds her to the window so that she can see the rain.

Though I didn't go far enough in math to understand the Weil conjectures, nevertheless I wonder, to what extent could I appreciate more about them? A bee in my bonnet, a dubious goal: maybe I could try to apprehend something of their flavor, I speculate, but at the same time I don't know what that would mean. What sort of apprehending would it be? I can only recall André admonishing his sister that explaining his work to a nonmathematician would resemble explaining a symphony to a deaf person. One has to resort to mere metaphor.

Still, I think, I could try to suss out the metaphor. I decide to look for a mathematician who might be willing to talk informally about the conjectures, and so I scan the website of the University of Texas math department. With a nervous prickle I realize that one of the professors at UT is someone I met long ago, at that math summer program—I wasn't sure what I was doing there, while his destiny was clear: he would be a mathematician. Back then he had the pale, soft look of a person who spent his days indoors; now, to judge by the photo on the website, he's a bit tanner and slimmer, but of course it's him, absolutely the same guy.

At the sight of the photo an interior seam splits, and it comes rushing back to me, how intimidated I was by them. The true math people, I mean, the wunderkinds, the superstars. This included my professors—I was terrified of all of them—and a number of my quote-unquote peers. Along with their myopic half smiles, their awkwardness, their naïveté, their sandals, their hacky sacks and ping-pong paddles, they had, or seemed to have, a solid confidence, an inner compass pointing directly to a university math department. They struck me as happily at home in that world, they had arrived, whereas any time I had reason to venture into the math department I would scurry in and out as quickly as I could. From those exceptional kids I detected (or at least imagined) some mix of snobbery and pity toward someone like me, smart enough to get by, but just the ordinary type of smart. Much as mathematics came with a democratic ideology, according to which it was a realm of rarefied knowledge open to anyone who wished to work her way along its paths, there also seemed to be an unstated but obvious hierarchy. If math to me was a dark place where I went groping around on my hands and knees, here were these other people with killer night vision who could see everything at once, go prancing from one topic to the next.

Weeks pass before I actually write to the mathematician. During this time I mention to a few people that I'm thinking of trying to interview a math professor but have been reluctant to contact him because I'm intimidated and also not entirely clear about what I'm asking, and they all say the same thing: I'm sure he'd be flattered to hear from an interested person outside his field, I'm sure he'd love to talk to you—but I'm not sure. In my mind I am that deaf person, nagging a composer to explain a symphony. One day as I'm driving I think I catch sight of the very man, that mathematician, on a bicycle, crossing a street up ahead, but I'm not close enough to get a good look.

At last I compose an interview request, not as an e-mail but as a document that I save to my computer. The next day I tinker with it, save it again. I revise it once more the following afternoon. It's a Friday when I paste the text into an e-mail and hit Send, and in that instant I'm already sure that I've made a mistake.

I keep checking, into the evening, for a reply, fully expecting him to put off an interview or turn me down outright. Nothing that day, nothing over the weekend. On Monday I suppose he's returned to work and will reply soon, but still nothing comes. Nothing all week. Nothing at all.

Never mind the possible explanations for this—an aggressive spam filter, or an in-box so jammed with other e-mails that mine was lost in the mix, or just the ordinary pileup of life that would outweigh a near-stranger's request, or some ordeal that would eclipse it entirely—I'm quick to interpret the lack of response as confirming the scorn I always suspected was there, disdain for the tourist.

A lot of Simone Weil's writing is awfully high in fiber, and it can be hard to digest, just as it is in her college essay on Descartes. Much as she critiqued her brother's mathematical abstractions, her prose is often quite abstract. She introduces ideas and relentlessly loops around them. At times her propositions have hardly any more meaning to me than the Pythagoreans' Justice = 4, only without the whimsical concision of Justice = 4. Yet her work is also sown with passages of wisdom and beauty, so that right when I'm tempted to start skimming, something will hook me, and I dog-ear the page so that I can return to it later.

I assume that I would experience Simone as I do her writing, that at times I would find her too earnest, an exhausting companion, saying the same thing over and over, obsessed with that day's obsession, yet at other times she'd be spirited and generous. An unexpected mischief would surface. This is the Simone who, having at last secured passage on a Swedish freighter bound for Liverpool, gathers the small group of fellow passengers every evening and, under an overcast sky that seems to reflect their collective foreboding, entertains them by telling folktales.

And here she is at the Liverpool clearing center, where she is detained for two and a half weeks while the British perform their background checks: she learns to play volleyball, of all things. Someone's put up a raggedy net on a dirt lot behind the cafeteria, and often she's the first one there. She comes dressed in baggy pants and a button-down shirt and a hat that she removes before playing. Aiming for the brick wall of the building, she tries to teach herself to serve. She almost knocks out a window.

Honestly I think I understand anyone else's dislike of math better than I understand whatever hold math has had on me. In response to the person who asks, "But what's the point of all this?," I don't have a good answer. In fact I sometimes *am* that person, can muster only a peevish respect for, let's say, Monsieur Pierre de Fermat, royal councillor at the parliament of Toulouse, who, when he wasn't engaged in the dreary business of regional administration, would turn his attention to another convoluted system of rules, searching for laws governing the whole numbers.

An uncredited portrait of Fermat, reprinted in a book I took out of the library, renders the great seventeenth-century amateur in an ample black cloak, with flowing dark curls and a pencil mustache. He looks, to me, rather fat and supercilious. In 1658, apparently, he presided over the trial of a defrocked priest and sentenced him to be burned to death. Meanwhile he laid out his results in letters to his contemporaries, to Mersenne, to Carcavi, to Pascal. Here's one: every prime number that can be expressed in the form $4k + 1$

can also be expressed uniquely as the sum of two squares, $a^2 + b^2$.

In college I was never tempted to take a course in number theory, which considers questions like this one, because it seemed to me too insular, a kind of numerical navel-gazing. I am pretty much unmoved by the fact that certain prime numbers can be written uniquely as the sums of two squares, and although now I believe that number theory is much more interesting than I gave it credit for, I imagine that many (if not most) humans feel about math the way I felt then about number theory. It's remote, it's not about anything but itself except by a series of happy accidents.

"Fermat lived temperately and quietly all his life, avoiding profitless disputes," claims the mathematician E. T. Bell in his book *Men of Mathematics*, a hoary classic of math history published in 1937. "We shall understand his even, scholarly life if we picture him as an affable man, not touchy or huffy under criticism," Bell asserts, and then in a confusing juxtaposition he adds that Fermat was "without pride, but having a certain vanity."

Humble, but haughty? Then again, E. T. Bell was paying tribute to his heroes. Over the years it has emerged that *Men of Mathematics* is not always the most reliable source.

A slight woman, plainly dressed, pushes open the door to an office in London. The boss thinks at first that this lady has come begging, that somewhere there are curly-haired pauper children to feed, but then she introduces herself, she's the new hire, Simone Weil, and still all he can think to do, since she seems on the verge of collapse, is to help her to the nearest chair. She slumps over herself. Her long black skirt bunches on the floor. He wonders whether she has fallen asleep, until she starts to speak. With hardly any preface she announces that her mission, of which the highest authorities have been made aware, is to drop herself into battle by means of a parachute. She has trained as a nurse, she adds.

The boss offers her a glass of water. She scans the room, tells him again that she has medical training.

For a few days the question of what to do with Simone hangs over the Free French office in London, until somebody finally says, well, she's a writer, let her write. They give her a small room with a desk. Various proposals have been made regarding how to reorganize the French government once the war ends, and they ask her to evaluate these schemes.

Far-fetched as the parachute idea seems to everyone else, she won't let it go, keeps mentioning it, and at the same time she works harder than the boss would've imagined anyone could, much less this woman. She writes day and night, page after page; it becomes obvious she's not just analyzing parliamentary structure. Having refused the offer of a typewriter, she composes everything in her careful, neat hand: well-formed letters that stand up straight and rather prettily on the lined page. It's as though she's taking dictation from a sourceless voice that speaks slowly, steadily, incessantly.

Rather than hand over her work in person she leaves it at the boss's door. Included in her typescripts are reflections about truth on a higher plane, about the idea of a good in the universe that is not just a human construct but an objective reality. The boss shows it to his assistant, who as he reads moves the paper farther and farther away from his face, wincing, and then asks, "But why doesn't she concentrate on something concrete?"

She's trying to learn to drive, they discover. She gets her hands on an aviation manual and studies it carefully. She obtains a parachutist's helmet.

The essence of powerful speech, according to Simone, lies in "bare simplicity of expression." Writing should be stripped-down and impersonal, thought scrubbed of any trace of the thinker. The greatest creations of Western culture, to her mind, are ones that achieve this transcendent anonymity: the *Iliad*, geometry, Gregorian chant.

Naturally, then, she doesn't insert herself directly into what she writes; her texts proceed by force of argument, unconcerned with style. Shortly before leaving France she passed off her notebooks to Thibon, suggesting to him that some of the material from them might be published under his name, since what mattered to her were the ideas they contained, not the fact that she was the one who wrote them down. Yet these notebooks, hundreds of pages of fragments, will eventually constitute a key part of her literary legacy—that is to say, the very pages from which she was ready to delete her name will, later on, help make her name. Unpublished during her lifetime, they're full of quotations and precepts, philosophical and religious and critical meditations, a working out of the principles of her own mental experience. They seem like notes for some hugely ambitious future book, a Bourbaki-like survey of existence as she knew it, starting from first principles.

For Simone, impersonality is not just a writerly ideal but an existential and religious one. She wants to bring about her own undoing. What she aspires to is a state of selfless perception, in which her mind wouldn't be limited to the data of her own senses. Rather, by suspending her own imagination and holding herself at attention, she thinks she might receive divine wisdom from the material world the way a blind person is informed by a cane: "May the whole universe, in relation to my body, be to me what to a blind man his stick is in relation to his hand. His sensitivity is no longer really in his hand but at the end of the stick." The universe would become an instrument. Through it she would know God.

The lecture room where André is installed that fall and winter is drab, and the acoustics are poor. It's as though the room's soundscape were partitioned into disjoint subsets, a well of his own words pooling around the lectern where he stands, and other, more agile noises floating above the students in eddying currents of their own murmurs and the timbre of the pipes. Every day but Friday, groups of army recruits are led into the room by their noncommissioned officers. The soldiers are silent and well-mannered young men for as long as they're in line, but as soon as they spread out and take their seats they revert to boyhood, unable to stop wiggling and stretching and whispering and snorting. Little did André ever think he'd miss the dull grinds from the year before, and yet these kids care even less for math. Really they don't have any use for it at all. They've been awaiting orders to deploy, and their commanders, searching to occupy them in the meantime, have dispatched them here to learn algebra and analytic geometry. It's make-work for both the troops and the colleges, which have plunged into poverty for lack of students.

They barely take note of the algebra that André is presenting—off the cuff, having not bothered to prepare a lesson. The interlocking machinery of simple equations, the formulas that delighted him at the age of nine, can't breach the hard shells of their brains. The army boys imitate his voice under their breath, then grow bolder and start to talk among themselves. One of them sneezes twice, and even that seems aggressive, such loud sneezing. André has to ask an officer to settle them down.

When it's calm, a boy raises his hand.

"But I don't understand what x is."

André smiles to himself. If only he could deliver a real response, convey something of the scope of the question. He could devote a lecture just to the history of mathematical symbolism, to Diophantus, Viète, Descartes.

A thousand images come to mind—himself as a boy with his algebra book, the half smile of his teacher Hadamard. Eveline as she was that summer in Finland. His prison cell. What is it? What is x?

He returns to this room, these boys, and starts to explain (again) the use of the variable.

———

O men of mathematics!

In the middle of watching the twenty-first online algebra lecture, I hit a wall, stricken with a sudden exasperation as much physical as it was mental, as though my head, my shoulders, part of my spine, were itching from the inside. While Professor Gross was elaborating on the Sylow theorems, as he was saying that "any two Sylow p-subgroups H and H' are conjugate," I became instantly tetchy, I couldn't take it any longer. Who cares? I am a midlife mother of two, I thought morosely, and this is the most pointless thing I could possibly be doing.

Though on second thought it was no more pointless than anything else I would be using the Internet to do. One day I'd opened YouTube and seen featured there, under the heading "Watch It Again," a row of five videos I'd recently viewed. In the middle of the row were stills from algebra lectures, a trio of Professor Gross action shots—Gross writing on the blackboard! Gross gesturing at the blackboard! Gross gazing at the blackboard!—and flanking those were a pair of 1980s music videos by New Edition.

That said, I paused the lecture.

At the same time it occurred to me that maybe the problem wasn't too much math but rather too little, a lack of immersion, since I only watched these lectures once or twice a week, if that. In the interim, I would forget about Sylow p-subgroups entirely (even as New Edition's "If It Isn't Love" stayed intractably stuck in my head).

"This is not like an everyday job which one can interrupt

any minute and resume again," said Gauss of math. "One always has to invest a lot of effort and has to have much free time to again bring everything to one's attention."

I didn't have the free time, the attention. It had been months since I started watching the lectures, and I was only a little ways past the midpoint of the semester, and it was far from clear to me what would be the merit in continuing with the class, other than whatever merit comes from finishing what you've started.

On a cloudy spring day, at a convent outside of London, a nun greets two young Frenchwomen, Simone and a friend, who have appeared at the gate. What can she do for them? Simone mumbles and trips over her words, and it takes some back and forth before the nun understands that they wish to camp on the grounds. The nun looks at the sky—thunderheads loom in the distance. Are you sure? she asks.

Well after vespers it begins to pour. Straining to see from her room's window, she can't make out anything and decides that the women must have left, but after adjusting her vision she finds their slick and collapsing tent. Sighing, she puts on her rubbers and runs out to them, urging them to come inside, as her umbrella pulls away from her and rain hits her slantwise, but only one of them agrees to follow her back to the building. The other, the one who asked permission to camp there, insists on staying in the tent, which is hardly any different from staying out under nothing at all, so bent up and abused and waterlogged is the canvas. Shivering wildly even as she refuses. The nun recognizes it, this love of misery, she's known fellow nuns who angle toward suffering like plants to the sun, but she pleads with the woman. You'll catch your death out here, she shouts. Catch my death! the woman repeats, as though she hasn't heard this turn of phrase before. Catch my death, she says again, and stays out there all night long.

My exasperation with the Sylow theorems echoed the way I'd begun to feel about math in general by the time I was a senior in college, as though I'd been squeezing myself into a container I no longer wanted to stay inside of, which was also the way I'd come to feel about my relationship with my boyfriend, even though these were good containers and had been very good to me and maybe I could've lived my whole life, another life, comfortably within them. But I grew restless, I threw all that away, just to see what I could see.

"There is a reality located outside the world, that is to say, outside space and time, outside man's mental universe, outside the entire domain that human faculties can reach," Simone writes while she's in London. "Corresponding to this reality, at the center of the human heart, is a longing for an absolute good, a longing that is always there and is never satisfied by any object in this world."

Beyond the quotidian real world and everything we know, an inaccessible goodness: for her, every search is a search for this, doomed to failure in the sense that these searches can't attain their end during this lifetime, but then again the quality of a life derives from the quality of its searching.

The mathematical conjecture, labeled as such, is a creature of the twentieth century; while there were always open problems and speculations and hypotheses, the spread of *conjectures* as a genre of math is a recent development. Barry Mazur—a Harvard mathematician I once took a class from but never dared speak to—calls these contemporary proposals "architectural conjectures," meaning they lay out the basis of a theory, a core set of expectations believed to be true, a foundation and a design that waits for others to come along and construct the rest of whatever is being built, maybe redeveloping the surrounding area in the process. The driving force behind all this conjecturing, Mazur writes, has been analogy, as in the case of the Weil conjectures. What the architects have been designing are bridges.

From André Weil, whose work I don't understand, I have nevertheless gained an indirect appreciation of number theory, one that begins with my picturing him in his prison cell, staking out his own colonies among the fields and functions, or later on in his basement study in Princeton (he joined the Institute for Advanced Study there in 1956), his cat perched on the edge of his desk, his family's footsteps sounding against the floors above him.

The closest I can really get to the Weil conjectures, as much as I can grasp, comes courtesy of our man of mathematics

Pierre de Fermat. In a letter written in October 1640 to one Bernard Frénicle de Bessy, an official at the French mint who had a gift for mental arithmetic, Fermat proposed what later became known as Fermat's little theorem—not to be confused with the more famous assertion known as Fermat's last theorem, though in both cases Fermat himself did not provide a proof: "I would send you the demonstration, if I did not fear its being too long," he informed Frénicle de Bessy. (Did he really know one? Or was this what Bell called a "certain hint of vanity"? The first proof was published almost a century later.)

This theorem is fairly straightforward. In brief: Consider a restricted set of numbers, these being the whole numbers less than some prime number p. If p were 5, for instance, we'd be talking about $\{0, 1, 2, 3, 4\}$. Any number larger than p is defined to be equivalent to one of the smaller ones, by taking the remainder when you divide that larger number by p. So 5 is equivalent to 0, 6 is equivalent to 1, 7 to 2, 8 to 3, 9 to 4, 10 to 0 again, and so on.

Fermat's little theorem states that in these realms $\{0, 1, 2, \ldots, p-1\}$, you can take any number a, raise it to the pth power, and get a back. Another way of saying this would be to say that all the numbers in our little domain are fixed points of the map that takes x to $x^{\wedge}p$. And this turns out to be a forerunner of the Weil conjectures, which hinge on an analogy between, on the one hand, counting the fixed points of continuous functions that relate points in a mathematical object called a topological space to other points in that space, and, on the other hand, counting solutions to systems of polynomial equations over finite fields of numbers.

Even Fermat's relatively simple theorem starts to grow hair when I try to lay it out in ordinary language, I realize, and it's hard to articulate why it's interesting without invoking more math. At the end of the day, why would it matter to a nonmathematician that André Weil figured out how to count solutions to polynomial equations in finite number fields? In one sense, it doesn't, not to me. I don't understand it well enough for it to matter. But at the same time there's a flicker of fascination, a door that cracks open just a sliver when I learn about these constructed realms and the relations within and among them, whether the realm is as simple as the numbers from 0 to $p-1$, or something too complicated for me to fathom. It's not so much the particular result as the intricate mesh of them that moves me: models nested within models, labyrinths built on top of labyrinths, the unlikely connections—the eros that André wrote about in his letter to Simone—in this mental universe.

And again I picture the prickly mathematician in his basement, surveying one section of his landscape while the cat looks on.

Simone lands in the hospital and never recovers.

Since leaving France she's eaten less and less, wanting to consume no more than what is rationed to children in France—a quantity concerning which she has no actual information. She makes up an amount, then restricts herself to less than that, in the same way that in the past she thought (mistakenly) that the working class was forced to live without heat and so refused to heat her own rooms.

Despairing, hardly eating, working herself to the bone. In April a friend finds her collapsed in her room and brings her to the hospital, where she is diagnosed with tuberculosis. She needs rest and food, they say. She spends three months in Middlesex Hospital, resting but not eating enough to recover. Her digestion is shot and she has no appetite; some days she's too weak to hold a spoon. A friend who comes to see her is appalled by her condition. Another is touched by what he perceives to be a spirit on the brink of releasing itself from the flesh. She implores everyone who visits not to inform her parents that she's in the hospital, and writes her former address on the letters she sends them, misrepresenting where she is. Her letters are one long lie full of tenderness, her friend and biographer Simone Pétrement would write.

The less she eats, the stronger her wish to take Communion becomes. In Simone's final months the Abbé de Naurois, chaplain of the Free French forces, makes three trips to see her. I picture him, no doubt wrongly, with a pasty but smooth complexion and a fine wool suit, the clever third son of an industrialist, let's say, his intelligence palpable enough for Simone to grab and shake. That is to say, they argue the way she argues with every man of the cloth. She claims she's only trying to find out whether he would consider her eligible to be baptized, but under that pretext she rattles on without listening to his responses, criticizing Catholic dogma, zeroing in on the church's doctrine of salvation and wanting to know precisely whom it includes or excludes.

The abbé sits next to the bed with his hands on his knees as this woman, this febrile figure in spectacles, half covered by a sheet, barely strong enough to move her legs, her arms even, binds him in a long, baffling chain of logic, then pauses and is silent for a while. Then she lets fly another knotty sentence.

The abbé interrupts, or tries to—her thinking is confused, he finds, contorted, swerving this way and that. "The acrobatics of a squirrel in a revolving cage," is how he'll one day describe her ruminations during those visits. He sits there, and as her flood of argument washes over him, he begins to doubt the very worth of the intellect, that it could spew this hairball of thought that, it seems to him, could only interfere with the spiritual contact she longs to experience.

Yet at the end of each meeting he blesses her, and she goes silent and is suddenly so gentle. Docile even, a wide-eyed young girl. Here is an extraordinarily pure and generous soul, he realizes. He will remember her that way, as a paragon of seeking.

The Weil siblings both undertook to translate into language something beyond words, beyond symbols, in Simone's case maybe beyond thought itself. I can only follow either of them so far, reading their words and making guesses as to what lay beyond articulation. Each had the run of an elaborate mental (or mental-spiritual) universe, each subjected perceptions to a ruthless accounting.

They thought their way into esoteric domains, found purpose in concentrated inquiry and likewise in the glimpse, the pursuit, the almost there, the exhilarations, the frustrations, of being partially shown and at the same time denied the dangling fruits of their searches.

———

Weeks or months, in wartime there's no telling how long it might take for a letter to cross the Atlantic, and so the correspondence between André and Simone is erratic. Their dialogue dwindles, and in what will turn out to be their last letters, they hardly know what to say.

Try to write us sometime and let us know how you are doing, André asks.

I haven't written until now because it's truly difficult to know what to write, Simone begins her reply, this on the heels of a period during which she seemed unable to stop writing, when about subjects other than herself.

London is full of fruit trees and flowers, she writes to André.

I have lately made the acquaintance of several charming young girls, she reports to her parents, omitting the fact that these girls are nurses in a hospital where she is a patient.

Here nothing new, André writes in July. Your niece is growing normally and continues to have a happy character. She seems to find life pleasant and agreeable.

A month later, he receives a telegram, out of the blue. YOUR SISTER DIED PEACEFULLY YESTERDAY, it says. SHE NEVER WANTED TO LET YOU KNOW.

She was thirty-four. "The deceased did kill and slay herself by refusing to eat while the balance of her mind was disturbed," reported the coroner. It's not entirely incorrect to say that she starved herself to death, but the full story is stranger than that, both more and less gruesome, or maybe it is better said that the details of self-starvation are not necessarily what we would imagine, or that we wouldn't truly be able to imagine self-starvation at all. What were her intentions? She may not have willed her death, yet it seems as though she might have been able to will her survival, or at least made more of an effort to survive.

The friends who visited her didn't believe that she wanted to die. Still she wrote, in a letter to another friend, "I am done, broken. Perhaps the object might be provisionally reassembled, but even this provisional reassembly can only be done by my parents"—her parents who had always swooped in to rescue her when she was on the brink but who were now stranded in New York, unaware of her condition and without visas.

She tried to eat, requested mashed potatoes prepared in a certain French way, consumed an egg yolk mixed with sherry, after that refused food for fear she could not tolerate it.

Blind man's stick, she wrote in her notebook.

To perceive one's own existence not as itself but as part of God's will.

A supernatural faculty.

Charity.

The eternal part of the soul feeds on hunger.

Nurses.

PART FIVE

II. SIMONE'S NOTEBOOKS ARE SENT TO her parents, who lug them from one country to another, all through their wartime migrations, from New York to Brazil, then to Switzerland, and then at last back to Paris, where the apartment on rue Auguste-Comte has been stripped of its furniture. All that remains is a large framed wall mirror, so tarnished it barely returns their reflections, not that they're inclined to look. They ease into the old rooms, now slowly molting a fur of dust and vacancy, everywhere shadows—unexpected traces, seen from the corner of an eye if never dead-on, of the other rooms they've taken and left since they departed these, the hotels and apartments and the camp in Morocco where Simone sat all day in that wooden chair.

There's a trail of dead leaves along a windowsill, a desiccated scrap of orange peel, a smell of kerosene. They heave open the casements, lift the limp clothing from their trunks. They buy beds and tables and lamps. And then, once everything is in order, to the extent that it can be in order, they start on the notebooks: every day Bernard and Selma, not quite ruined, sit across from each other at the wooden dining table and copy their daughter's thoughts, line by line, from her original notebooks into a series of blank accounting ledgers. Two bent heads, two pairs of eyeglasses. They follow along with their fingers, mouth her words as they write.

Having outlived her, having to swallow daily this fact that won't stay down, everywhere they go they find the same bleak streets, the same wrecking balls of memory. Maybe they wouldn't even remember whose idea it was to transcribe the notebooks. They're so steady in their efforts that for a long time their granddaughter Sylvie, when she comes to visit, concludes that this work of copying is their job. And really she's right, it *is* their job, though no one has hired them to do it. They've become their daughter's most diligent students, reading her, glimpsing her between the lines.

(How she went stomping through doorways.)

They keep the original notebooks and the copies in a cabinet in the living room, where other people might keep linens or vases or playing cards.

(A stub of pencil between her lips as she read the newspaper. Or at work in the other room, calling to them that she would come out to eat in just a moment. *J'arrive!*)

But would you, could anyone, want this? That in the event of your death your parents would slowly, laboriously copy the contents of your notebooks into other notebooks?

(Her face when she listened to music.)

(Her little hands.)

Day after day.

Philolaus of Croton was said to have been the first to reveal in writing the beliefs of the Pythagoreans. He was also said to have conjectured the existence of a planet no one had seen: since the number ten was, in the Pythagorean view, the most perfect number, and only nine heavenly bodies were known (the earth, the moon, the sun, and six planets), he postulated a tenth one, an invisible planet. A counter-earth.

According to one story his work was plagiarized, by none other than Plato. There is also a theory that many of the surviving fragments of his writing may not be his writing but instead a forgery, produced much later by someone familiar with Aristotle's account of Pythagorean thought.

One way or another, posterity has linked him to stolen ideas.

Counter-children, counter-trees, counter-courtyards, counter-dishes, counter-dogs?

Simone would allude to him repeatedly in her last notebooks, that is to say, she invoked the "Philolaus" who authored the texts, who may or may not have been the true Philolaus. She had a way in the notebooks of cycling through what she knew, returning again and again to certain lodestones of her thinking. Passages from the Bible, from the ancient Greeks, fragments by which she recalled herself.

Sometimes she just wrote the name "Philolaus"—an invocation of the idea (stolen or not) that mathematical truth was, in those ancient times, inseparable from spiritual truth, that mathematics was a bridge to the divine.

————

"And all things that can be known contain number," wrote Philolaus, or the imposter writing as Philolaus. "Without this nothing can be thought or known."

She kept turning over math in her mind. In those late writings there are not only notes on the history of algebra but pages of actual calculations in trigonometry and combinatorics. At the same time, she considered math from the outside, wrote of math as a model of certainty, math as an image of divine things.

She repeated to herself what she'd written privately to her brother—that contemporary mathematicians had roamed too far off course. The reign of algebra, she would say, is like the reign of finance: Just as money has gummed up the relationship between work and its products, so math too has become divorced from the material world. "The relation of the sign to the thing being signified is being destroyed, the game of exchanges between signs is being multiplied of itself and for itself," she wrote.

Ultimately she wanted to think her way past thinking. For her, the ultimate goal, the thing for which math trains us, is to surpass the part of our brain that does math, to transcend reason, even to transcend time.

"One needs to have traversed the perpetual duration of time within a finite period of time," she wrote. "In order that this contradiction may be possible it is necessary that the part of the soul which is on the level of time—the part that reasons discursively and measures—should be destroyed."

Logic sometimes makes monsters, Poincaré said.

In the final chapter of his memoir, André tells of receiving the cable that announced his sister's death, a telegram that would forever remain, as he put it, etched in his mind. All he says of his reaction is that it was suppressed.

"How can I describe my grief?" he wrote. "But I did not have the luxury of indulging it; it was up to me to inform my parents, and I did not feel equal to the task."

It could be an accident of timing that the memoir, though he wrote it late in life, concludes soon after Simone's death, that the years of his apprenticeship, as he construed it, happened to roughly coincide with the years she was alive. Or not an accident and yet not the sole reason he would've bracketed his memories in that way: it was the sum of everything he experienced over the course of the war—prison, the military, emigration, the birth of his first child, and the death of his sister—that caused him to divide his life into before and after.

Still, I wonder whether his very mode of remembering might have altered after he lost her, once his first and best witness was gone. Decades later, responding to an interviewer who asked him why he had confined his autobiography to his first forty years, he said, "I had no story to tell about my life after that."

No story, even though after that, in 1946, he and Eveline had a second daughter, Nicolette, and after that, in 1949, he published a paper, "Numbers of Solutions of Equations in Finite Fields," laying out what would become known as the Weil conjectures. Not just an achievement but a landmark, setting a course for algebraic geometry to follow for the rest of the century. (A fresh set of equipment had to be invented in order to verify the conjectures—the proofs were assembled, over time, by Bernard Dwork, then by Alexander Grothendieck, and finally by Pierre Deligne.)

Yet he had nothing more to say about himself and sneered at the "very boring" autobiographies some of his peers had written, which seemed to him mere litanies of their academic appointments and the theorems they'd proved.

Simone was "naturally bright and full of mirth," André would write, "and she retained her sense of humor even when the world had added on a layer of inexorable sadness." As adults, he noted, they'd had few serious conversations.

"But if the joys and sorrows of her adolescence were never known to me at all, if her behavior later on often struck me (and probably for good cause) as flying in the face of common sense, still we remained always close enough to one another so that nothing about her really came as a surprise to me—with the sole exception of her death. This I did not expect, for I confess that I had thought her indestructible."

———

When does the story of a life end? Simone went on to a post-humous existence that she wouldn't have anticipated: she became famous, an intellectual celebrity in France and abroad, her work published and translated and admired by the likes of T. S. Eliot, Mary McCarthy, Albert Camus, Elizabeth Hardwick, Czesław Miłosz, Susan Sontag, and Iris Murdoch. Having written so relentlessly and died so young, she acquired, after death, the burnish of genius cut short, an Elliott Smith for the *Partisan Review* set. Her fans were intellectuals unmoored from tradition and shaken by the atrocities of war; maybe she spoke to a desire for some (necessarily cryptic, necessarily tragic) sense of what it all meant, a wish, even as they distanced themselves from the religions of their parents, to conceive of a philosophically respectable spiritual life, to rehabilitate the idea of the soul.

Not everyone was deferential. "The life of this remarkable woman still intrigues me while much of what she writes, naturally, is ridiculous to me," pronounced Flannery O'Connor, in a letter to a friend who'd sent her a collection of Simone's writings. "Her life is almost a perfect blending of the Comic and the Terrible, which two things may be opposite sides of the same coin."

Simone was more often quasi-canonized as a kind of genius or dismissed as a nutjob than she was recognized as something in between, as human. I suppose she herself bears some of the blame for this, determined as she was to detach herself as much as possible from ordinary personhood.

Outside of mathematical circles André was never known

in his adopted country. When an English translation of his memoir appeared in 1992, it was reviewed only in scientific publications. As a public matter, his story ended well before his death, while Simone's demise got the ball rolling. In this century, though, you might say there's been, if not a reversal, then a leveling: Simone's fame and influence have dwindled, while her brother's mathematical discoveries remain in place, struts supporting later advances made by others.

Sylvie Weil also published a memoir, *Chez les Weil*, about her father and her aunt and her own life in the shadow of all that genius. She writes of meeting a man who'd known Simone in London during the war and who remembered her, in specific and convincing detail, as fragile and tired and shy and isolated—which came as a shock to Sylvie, so different was his description from the aunt whom her father and grandparents had remembered, Simone as indestructible, as a force of nature.

The same man was certain that Simone had known, in 1942, of the deportation of Jews to Nazi camps. What has bewildered and troubled many people, her admirers and critics both, is that she never directly mentioned it, not once in all the reams of writing that spilled out of her before she died.

12.

WHAT IS IT ABOUT THESE BYGONE thinkers, these dead mathematicians, that captures me? A fond reverence muddled with a strain of muted pity, a distant, tainted love: I can't exactly name this feeling and so keep resorting to their stories; I'm still trying to describe it or at least circumscribe it, and so (with apologies) I'll indulge in one more digression.

In January 1954, a graduate student at the University of Tokyo went to the library to look for *Mathematische Annalen*, volume 124, only to find that *Mathematische Annalen*, volume 124, had been checked out weeks earlier by another graduate student, whom he knew in passing. The first student, Goro Shimura, wrote to the second one, Yutaka Taniyama, to ask whether Taniyama might return *Mathematische Annalen*, volume 124, so he could read an article that described a theory of complex multiplication. Taniyama sent his reply by postcard, addressed from the town of Kisai, where his parents lived. It appeared that the two of them were working on the same problem, Taniyama wrote; maybe they could talk sometime.

They were drawn to algebraic geometry and so to the work of André Weil, then teaching at the University of Chicago. Not long before writing to Taniyama to ask for *Mathematische Annalen*, volume 124, Shimura had sent a manuscript to Weil, in Chicago. As an undergraduate Taniyama had read Weil's *Foundations of Algebraic Geometry* as well as some of his papers. Both students had found little to admire in the older generation of Japanese mathematicians, whom they deemed full of themselves and all too prone to offering useless remarks in the guise of advice, and so instead they learned from each other and from distant idols.

At that time you could see, from the center of Tokyo, the crest of Mount Fuji some seventy miles to the west, a view that pollution would eventually obscure. Taniyama lived on the second floor of a run-down wooden structure wishfully named Villa Tranquil Mountains, though in reality it sat on a narrow urban street and bore no resemblance to a villa. A sort of a ramshackle dormitory, it had, on each of its two floors, twelve tiny dwellings—small rooms with a sink attached to one wall—and a shared toilet. Baths were taken at a nearby public bathhouse.

Taniyama, as Shimura would later remember him, had few interests outside of mathematics. He wore a shimmery suit made from fabric his father had bought cheaply; he ate whale meat and tongue stew at restaurants. He often left his shoes untied. He would work late into the night, sometimes until dawn. I picture the young mathematician at his desk, after midnight, in an undershirt and iridescent pants, hardly aware that he is picking a bit of blubbery gristle from his teeth. He grapples with fleeting modular forms and elliptic functions, while behind the thin walls of the surrounding miniature apartments, his neighbors work troughs into their pillows and dream of childhood friends or long journeys or snow.

Shimura was an early riser. He worked in the mornings, and often in the afternoons he would make the trip from his university office to Villa Tranquil Mountains. If Taniyama was awake, the two would huddle in his room and riff on the properties of elliptic curves over algebraic fields. If Taniyama was asleep, Shimura would return home and record the fact in a diary he kept with the precision of a captain's log. For most if not all of any given twenty-four-hour period, at least one of them would be up, and so together they could be a kind of round-the-clock mathematician, able to labor at a problem continuously.

Age is just a number, they say. Years are also just numbers—if at times they seem unjust, succeeding one another with such brutal linearity. One pleasure in writing arises from the illusion of holding those numbers in your hand, stopping time or running time backward or hopping back and forth from one year to another, making a game of it, huddled in an imaginary tent pretending to be a kid pretending to go camping.

Math affords a different kind of timelessness. The mathematician manipulates systems of abstract objects that seem to exist (in whatever way they exist) outside of time; she plays in a field of the forever true. And becomes, habitually, lost in thought. She looks up from her desk and notices that the sun has set.

Think of an ordinary curve, a looping line drawn on a sheet of paper, as a subset of all possible points on the paper. Now generalize that idea to other spaces, other dimensions: conceptual ripples in conceptual dark seas. Think of families of curves and of ways to bundle them together. Think of the path from Shimura's department office to Villa Tranquil Mountains. Think of the twisting course of the long walks Shimura and Taniyama take through the streets of Tokyo.

The two young men are at work on a series of tramways, erecting them section by section. One of them rises promptly, reports to the worksite, dons his hard hat, measures the existing track, and steadies the girders underneath it. The other, late at night, sketches unexpected new routes.

Some afternoons they meander about the botanical garden together, talking about Riemann surfaces, eyeing women as they pass. Or they visit a Shinto shrine where "oracles" are sold, fortunes printed on small pieces of paper. The early riser learns that he will one day have children. The late-nighter unfolds his paper, reads what is written there, folds it back up, and attaches it to a chain-link fence where hundreds of other paper fortunes flap against the metal—this being the place where unwelcome predictions are discarded.

Age is just a number, but your days are numbered. When Simone Weil died she was thirty-four. When André Weil died he was ninety-two.

When Yutaka Taniyama died he was thirty-one. As I write this, Goro Shimura is nearing ninety. In 1989, when he was in his early seventies, he published a beautiful essay in the *Bulletin of the London Mathematical Society*, called "Yutaka Taniyama and His Time: Very Personal Recollections."

In 1955, André agrees to attend the International Symposium on Algebraic Number Theory, held in Tokyo and Nikko. Weil himself in Tokyo! Lately a mood of ambition and optimism has spread through the whole country, one embodied by the young mathematicians, Taniyama and Shimura and a handful of others, who await André like they would a prophet. He arrives in the city several weeks before the beginning of the conference, and once they know he's nearby, they grow impatient, joking anxiously with one another.

It's the middle of August. They have a chance to shake his hand at a university reception, which doesn't exactly change anything for them, but then again the fact that Weil is in town changes everything, the sounds and sights of their city altered. From the outside everything comes at them more acutely; on the inside everything whirls, stops, whirls in the other direction.

One day Shimura is summoned to the Prince Hotel Annex, where Weil is staying. The professor wishes to meet with him and discuss his work, the secretary of the mathematics department informs him over the phone. Young Shimura wears a jacket and tie but finds Weil, standing in the lobby, in tan slacks and an open shirt. They speak English, rather formally, as they wander over to a small courtyard and sit down. A shiver vibrates through the younger man's body, for here he is in a private meeting with the mighty Weil, the name on all those papers and books, now attached to a short bespectacled Frenchman, who directs a young woman to bring them tea and cake.

No sooner has she turned her back than Weil hits Shimura with questions about his research. Shimura does his best to answer, though he's sweating in his jacket and his mouth is growing thick with foreign constructions. Is he actually listening to any of this, Shimura wonders, as the older man starts to scribble formulas on hotel stationery. Weil interrupts, then interrupts himself by coughing, then stands and begins to pace, marching toward the other end of the courtyard, where dogwood branches paw over the top of an iron gate.

He comes and goes, back and forth, and soon enough he's the only one talking, and talking some more, trying to pack Shimura's head full of his ideas, though Shimura's head is already at capacity. After the woman comes back with his order, Weil devours a large piece of cake. Shimura has lost his appetite.

Once the conference starts, the young mathematicians make prank calls to one another in which they pretend to be Weil. They'll call someone up and imitate his voice and try to speak English with a French accent. "Hello, zis is Weil," they'll say, in their Japanese accents.

In his research Taniyama took on a thorny open problem, and he would characterize his efforts to untangle it as "hard fighting" against difficulties and a "bitter struggle" of trial and error. Indeed, any significant mathematical undertaking, he said, was a matter of hard fighting and bitter struggle.

Shimura noticed something else about the way his friend worked: Taniyama was "gifted with the special capability of making many mistakes," he wrote, "mostly in the right direction."

A notion that turns out to be wrong might still point the way forward, provided it's wrong in the right way. "I envied him for this, and tried in vain to imitate him," Shimura continued, "but found it quite difficult to make good mistakes."

Corazonada is a Spanish word for *hunch*, and I like how it implicates the heart (*corazón*) in our intuitions, not just the bent spine that I'm reminded of by the English. At the conference Taniyama discloses his emerging hunch, the heart of his work so far, to Weil, who is eager to discuss it.

WEIL: Do you think all elliptic functions are uniformized by modular functions?

TANIYAMA: Modular functions alone will not be enough.

I think other special types of automorphic functions are necessary.

Et cetera.

Fluttering about the conference is a crew of pretty young hostesses, there to help the foreign visitors. André mentions one of them, Momoko-san, in his letters home, and his daughters are delighted to imagine her as a kind of doll come to life, while Eveline is not so delighted.

There's a passage in Proust's *In Search of Lost Time* in which the narrator rereads a letter from a lover who has died, and he feels, fleetingly, a joyful surge of expectation, as though he'd been sent back to the time when he first read the letter, while his beloved was still alive. The narrator then reflects upon how, remembering episodes from the past, we may find ourselves suddenly disoriented in time, inhabiting a past moment as though it's our current reality, so that we briefly, delusionally look forward to a future that's already gone by. We anticipate prospects that already came to pass (or didn't) years ago. "The illusion swiftly dies," Proust writes, "but for a second we felt ourselves driven forward once more: such is the cruelty of memory."

I wonder whether Shimura experienced this phenomenon as he wrote his essay, whether in remembering Taniyama he might've slipped into looking forward to seeing Taniyama again, to walking through the botanical garden with him, to discussing math problems—only to fall, again and again, back into the present.

Of course I don't know, but his essay from the *Bulletin of the London Mathematical Society* is suffused with a melancholy that gives way, at its close, to grief for his lost friend.

The idea that Taniyama and Weil bandied about at the conference would crystallize as a conjecture, one that Shimura made precise and conveyed to Weil, who articulated it in a paper that failed to mention either Shimura or Taniyama. This conjecture would be called, at first, the Weil conjecture (distinct from his 1949 Weil conjectures), yet as details of its genesis became more widely known, the name was modified to the Taniyama-Weil conjecture, then the Shimura-Taniyama-Weil conjecture, or the Taniyama-Shimura-Weil conjecture; if you were hostile to Weil you called it the Taniyama-Shimura conjecture or the Shimura-Taniyama conjecture; while if you enjoyed provoking Shimura you might still call it the Taniyama-Weil conjecture. One mathematician, Michael Harris, attempted to sidestep the whole mess by calling it "the conjecture associated with the names of" Taniyama, Shimura, and Weil. Nonetheless, Harris wrote in a blog post, he and a colleague who'd also adopted this construction "walked into an avalanche"—of criticism, presumably. Now it is not uncommon to drop all the names and refer to it as "the modularity conjecture for elliptic curves."

It says: every elliptic curve defined over the rational numbers is a modular form. Which is over my head, but the crux of it, I'm told, is another unlikely link between topology and number theory.

In 1957, Shimura left to spend a year in France. Shortly before his departure, he and Taniyama attended a dinner party given by a young woman named Misako Suzuki, who made fun of Taniyama because he hardly spoke during the meal. Not long after that she and Taniyama became engaged and signed a lease on an apartment. Shimura would call her a "typically pleasant girl from a typically upper middle-class family," whom he never got to know well.

It wasn't Shimura, by then in France, but the superintendent of Villa Tranquil Mountains who visited the apartment on the morning of November 17, 1958, five days after Taniyama's thirty-first birthday, and found him dead. There is no mention in Shimura's article of how Taniyama ended his life, only that he did.

Taniyama left a three-page note on his desk. "Until yesterday, I had no definite intention of killing myself," it read. "But more than a few must have noticed lately that I have been tired both physically and mentally. As to the cause of my suicide, I don't quite understand it myself, but it is not the result of a particular incident, nor of a specific matter. Merely may I say, I am in the frame of mind that I lost confidence in my future."

His friends and colleagues, Shimura writes, were utterly perplexed, overcome by shock and sadness. A brilliant career, a fiancée with whom he'd recently been furnishing a new apartment. There was so much ahead of him, in that future he decided all of a sudden to abort. Misako Suzuki killed herself two weeks later.

When he returned from France, Shimura wrote, spring was quickly passing. Though he'd only been away for a year and a half, and the city streets were as flashy and bustling as ever, he knew that an era of his life, "the years of turbulence" as he called them, had ended.

13.

AT THE TOKYO AIRPORT, A STOOPED and wizened André Weil is following Sylvie away from the baggage claim, and she in turn is following a driver, a young man in a cap who met their plane and is guiding them to his car.

Give me your hand, I cannot see anything, André demands. With my poor eyesight, how am I expected to manage?

His eyes, soupy with cataracts, are failing him, everything is failing him, his ears are fitted with hearing aids, his hips are made of plastic. His skin is loose and speckled.

Rather than take the hand that Sylvie offers him, André seizes her forearm and bears down, and they wobble forward, Sylvie seesawing between her father on one side and her shoulder bag on the other. Lowering him into the car is another delicate operation. He snaps at her and at the driver all the while. At last he lands, and the seat itself breathes a sigh. Neither André nor Sylvie is in any hurry to reach the hotel, where they'll have to do it all in reverse.

It's 1994, and Eveline has been dead for eight years. André has been spending every spring in Paris, rattling around the apartment on rue Auguste-Comte, alone much of the time, though Sylvie, who goes back and forth between

New York and Paris, is often in town. Mathematics has passed him by and he's left with memories of it, of lines of poetry, of his school pranks. A certain wistfulness regarding the Riemann hypothesis, a profound conjecture that he pursued (and proved a special case of) but never managed to slay. Once or twice a year someone thinks of him and publishes a laudatory article or presents him with an award. Then there are those who persist in asking to interview him about his sister, though when it comes to Simone, he's said all he has to say.

After he learned that he'd won the Kyoto Prize, Sylvie tried her best to discourage him from traveling to Japan for the ceremony, but in the end she relented and agreed to go with him.

On the plane to Tokyo he drifted into a reverie about Momoko-san, that lovely girl from decades ago, unfailingly generous . . . My God, he thought as he descended from his dream, she must be at least sixty now. And what would she think if she could see him, in this changed form? Suddenly irritated, he told his daughter to ring the call button, he wanted a coffee.

They don't venture far from their hotel until the next morning, when they board a train to Kyoto. As Sylvie is dressed all in black, André cracks that she is properly attired for his professional funeral. He then stares at a timetable, though he can't read the Japanese, until his eyes start to close. Much as he insisted on coming to receive his prize in person, now the thought of sitting through the ceremony enervates him. Lately he has a way of slowing down and then reviving, unpredictably; a strong smell or a flash of light will cause the unoiled gears of his mind to turn more quickly, and he becomes himself again, the learned professor. It's as if he's been given some temporary drug that makes him ten years younger, and for thirty minutes, an hour, until it wears off, he'll be witty, erudite, expansive.

At their next hotel, a retinue of local mathematicians and interpreters meets them at the door and welcomes them into the lobby—a giant clamshell of medicine-pink marble. André takes it all in, not just the marble but the gilded furniture, the heavy drapes, the sweeping staircase, the uniformed women at the reception desk, and in that moment Simone returns, it's Sylvie whose arm he's clutching but he can feel Simone there too, and his outrage is also hers, he's protesting on her behalf.

This is out of the question! he bellows, then initiates the arduous procedure of turning himself back toward the doors. Absolutely not, he says. I could not possibly linger for even an hour in an establishment for nouveau riche businessmen such as this.

He squeezes his daughter's arm even more tightly. His hosts smile, obviously panicked. One of them goes straight for a telephone closet. But Simone—not Simone, Sylvie—stops him.

Yes, it's ridiculous, she murmurs back, very crass, but I would find it so amusing to stay in an overpriced hotel. As an anthropological study.

A girl again, asking him for chocolate.

Please?

Later on, considering the menu of the hotel restaurant, Sylvie wonders whether it was the right tactic. They can't afford to eat there, so she asks a receptionist to recommend something more modest, which initiates an excruciating round of giggling and bowing and missed signals, then at last the drawing of a small, perfect map, to a place across the river. Father and daughter creep down a dark, narrow street of red lanterns. He reminisces about his past trips to Japan, and for a spell she is reminded of walks they took when she was a girl, her father's voice the very sound of security. The restaurant, though, seems as dark as the street, and he squints and scowls and is displeased to be told the seating is on the floor. First he has to get himself all the way down there, and then it's uncomfortable for him to sit like that, and although the food costs less than at the hotel restaurant, the prices are still out of reach. They dine on soup and rice.

He refuses to bow to anyone, lashes out at the waiter.

Would it be so hard for him to be polite? Sylvie asks.

May I remind you, he says, that your own talents would never get you an invitation such as the one that has brought us here.

In the silence that follows, she thinks of how he used to send her long letters during her student days in Paris, typically full of advice about her studies, though some of them were made up entirely of airline and ocean-liner schedules, which he'd copied so that she could plan her vacation trips to Princeton. When she came home, he would be holed up in his office in the basement, not to be interrupted, the eminent mathematician at his desk. Only the cat was allowed to keep him company.

They struggle back to the hotel and do it all over again the next day.

The idea that math is immortal, that its discoveries accumulate over time but that its truths are outside of time, is implicit in its everyday language of theorem and proof, all those statements made in the eternal present tense. But how can that be? How can math be timeless even as everything that underlies it—the historically specific ways that concepts are described, manipulated, and proved—shifts over the years?

Proof, that seeming ironclad warranty, is at the end of the day a rhetorical device, a method of persuading others of your conclusion. Proof in itself is hardly immune to history. It has evolved over the centuries, finding different means of expression and adhering to different standards of rigor.

But then again I'm not going to sit here and say that math is *not* timeless.

L.E.J. Brouwer, he of the nudism and the fixed-point theorem, had a change of heart regarding proofs. Influenced by his philosophical investigations—beginning in the 1910s, he argued that math was a mental construct independent of language—he decided to reject the method of proof by contradiction, upon which a vast number of math theorems rely, including his own fixed-point theorem. This method, a.k.a. reductio ad absurdum, depends on what Aristotle termed the law of excluded middle: something is either true or not true. You assume a proposition is not true, and if that leads to an absurd or self-contradicting consequence, then you conclude that the proposition must be true.

There was something unsatisfying about that kind of argument, Brouwer came to believe. He wished to establish math on a purely constructive basis. Which meant that if you wanted to prove that something (such as a fixed point) exists, you couldn't merely show that its nonexistence would imply the impossible, you had to give a prescription for somehow finding the thing.

His program attracted a few followers, but in the main, proof by contradiction lives on. And proofs have become even more oblique; nowadays it's not uncommon for them to run to hundreds of pages. Some require the verification of so many different cases that you need a computer to complete the task, so that the role of the human is to write the code rather than the proof.

Might there be some kind of wiggle room between true and false; in other words, can something not false be, at the same time, not quite true? Logicians have their undecidable propositions and what they call many-valued logics, while if we're talking about ordinary life, then easily yes, this is the case for all sorts of exaggerations and understatements, not to mention innumerable "I love you"s uttered in fraught circumstances. And this can pertain to ideas too, can't it? To conjectures that will never be proved? Maybe even to the idea that math is timeless.

I imagine the ghost of André Weil would have something acid to say about all this, that he would disdain my having ventured to write about him. In his later years he was clear about which people he thought were qualified to tell the story of mathematics and its pioneers, and those were mathematicians themselves—if not active mathematicians, then people of "better than average mathematical talent" who were in close contact with active mathematicians. Their proper purpose, moreover, was to aid working mathematicians in their research, insofar as revisiting the classics could inspire further advances. During his time as a member of Bourbaki, Weil had contributed historical notes to *Éléments de mathématique*, and when he neared retirement age, recognizing that his mathematical creativity had diminished, he took up the writing of history in earnest.

Around this same time—the 1970s—he assailed at least two historians of math who weren't trained as mathematicians. The first was a Princeton professor and the author of a book about Fermat, which Weil demolished in a review for a math journal; the second was a Romanian-born Israeli historian named Sabetai Unguru who could be every bit as vicious as Weil. Late in the decade, Unguru and several mathematicians threw punches back and forth in a forum that I'd guess was normally rather more genteel: a scholarly journal called *Archive for History of Exact Sciences*.

Mathematicians who resorted to writing history after they'd become "professionally sterile" in their own fields, proclaimed Unguru, would never see past their own biases; they couldn't help turning their chronicles into oversimplified narratives of progress. Men like Weil, in other words, were sorry has-beens and bad historians. Weil counterattacked, describing Unguru and his ilk as parasites, pretenders equally ignorant of history and mathematics.

The mathematicians' view of the past tended to be slanted

by, and served to reinforce, the Platonic ideal of math as an eternal structure somehow immune to historical forces, Unguru wrote. I think that was probably a fair criticism of how Weil regarded math's history, however low the potshots at older mathematicians.

But he always relished a good fight. After Weil died, Shimura would remember an incident from 1957 or 1958, when they both had positions at the Centre National de la Recherche Scientifique in Paris—Weil on sabbatical from the University of Chicago, Shimura a visiting researcher. One day he sought out Weil in his office to deliver a message. At the door of the office Shimura heard Weil and another man yelling at each other. Shimura knocked, the shouting stopped, the door opened, and Weil introduced him to his shouting partner, a visiting mathematician from Johns Hopkins. When Shimura left, Weil and the other man started up again. Shimura went on to the library, where he spent about half an hour. Upon his return he could hear that Weil and the other man were shouting still.

In what way do geometrical demonstrations in the books of Euclid belong in the same category as byzantine proofs devised by twenty-first-century computers? Maybe we could think about mathematics, the historian Moritz Epple has suggested, by analogy with music. If we hear a musical composition from centuries ago performed, we identify it as the same piece of music, although everything but the notes—

the instruments, the interpretation, the context, our whole manner of listening—has changed.

An audience with a princess. A dull bus ride to look at a temple, marvelously restored, they tell him, though all he can make out is that the building has been smothered in gold leaf like the girl in that James Bond movie. An interminable reception, where the champagne is too cold and he can't locate a chair.

André recalls a story about his contemporary Paul Erdős (a childlike vagabond, too fond of amphetamines, about whom there are many stories): that he once attended a party along with some colleagues at the house of another mathematician, and when they were well into the evening the others realized that nobody had seen Erdős for the last hour or so. A search began, and they discovered him upstairs, talking with the host's blind father, who had been sitting alone in a room when Erdős came upon him and decided to keep him company.

Now André is the old man, not blind but weary of looking, and though he and Erdős were never close, what he wouldn't give for his companionship just then.

If only. If only he'd cracked the Riemann hypothesis. If only Eveline were here.

Then the awards ceremony, inside a grand hall where women's bright kimonos stream like fireworks through a sky of tuxedos. A multitude of dignitaries and translators and assistants and photographers navigate to and fro. The floodlights, they tell him, are for the sake of the photographs.

They are blinding me, he says, and someone digs up a pair of huge sunglasses, which he puts on over his other glasses.

The prize has been granted to three men: André, a hulking American chemist, and the film director Akira Kurosawa, escorted by his own daughter, who wears a suit even pinker than the hotel marble. Kurosawa and his daughter are a head taller than André and Sylvie, and elegant, at ease in their own country, seeming to preside over the table where they are all sitting, even over the ceremony itself, where the director is the crowd favorite, at least for the half of the crowd that stays awake. Maybe that's why, at the reception afterward, André leans over to Kurosawa and says, "I have a great advantage over you. I can love and admire your work, but you cannot love and admire my work."

Suspended from their necks by thick blue ribbons are gold medals inlaid with synthetic jewels, sapphires and emeralds and rubies. In a photograph taken of André at the event, Sylvie will later write, he looks like an entry at a livestock fair—like a prizewinning pig or cow.

On their final night in Japan, André and Sylvie dine with a long-ago colleague, Shokichi Iyanaga, and a Mr. and Mrs. Satake who'd been helping the Weils in Tokyo. After dinner, they wish him a safe trip and hope there will be an opportunity to invite him back to their country. André moves his jaw a bit without speaking. Iyanaga, who is eighty-eight, the same age as André, waits for him, remembering how he used to pounce before Iyanaga had finished a sentence.

At last André says, "The next time perhaps in another world."

14.

THERE ARE TWO STYLES IN MATHE-matics, said Alexander Grothendieck, a titan of twentieth-century math. Picture a theorem as a hard nut, the mathematician's task to open it. One way would be to hit it with a hammer and chisel until it cracks, but another way, and this was his preferred way, was to sink the nut into water. "From time to time you rub so the liquid penetrates better, and otherwise you let time pass," he wrote. Eventually the nut opens easily, practically on its own. Math advancing through a series of imperceptible chemical reactions.

Or, he added, you could think of a sea rising, washing over hard earth until it softens.

It was Grothendieck who softened the shell of one part of the Weil conjectures by developing the right sort of mix-ture in which to dissolve them. Where Weil was fastidious and distrustful of big machinery, Grothendieck was the field's abstract expressionist, an otherworldly, romantic figure who emerged from a lonesome disaster of a childhood and made a haven of mathematics, mounting large, revolutionary canvases, only to resign at forty-two and return to a life of isolation. Another mathematician would compare him to

Simone Weil, noting that "his life was burned by the fire of the spirit." Like her, he was an ascetic, and like her, he believed that math was an attribute of God.

André Weil wasn't an Alexander Grothendieck (who left math to become a hermit), or a Georg Cantor (who was in and out of asylums), or a John Nash (a.k.a. the guy from *A Beautiful Mind*), not one of those troubled men who seem to arouse the most curiosity from outside the field and who contribute to the image of the great mathematician as an unhinged genius. André was a great mathematician who also happened to be sane—and irascible, prank-loving, imperious—a married father of two who lived a long and productive life.

But the more I learn about him and his sister, the more I begin to wonder whether his sanity somehow implicated his sister's extremity, whether in the Weil family, the two roles were divided between them: he would be the great mathematician, and she would come unhinged.

In Kyoto, André Weil gave an acceptance speech. I picture, at the podium, the chassis of a human thinker. A cranium in glasses, the mound of his forehead made larger by the retreat of his hairline, the rest of his face thinner than ever. In his tuxedo, with his little black bat of a bow tie.

He'd been asked to give a lecture of a personal rather than a technical character, he said. And so he recalled his childhood, the early love for math that had possessed him—when it came to his career, he said, "There was no choice on my part"—his great luck in being let into Hadamard's seminar, his travels. Then he spoke of Bourbaki, how it came about and conducted its business, even though by then the grandest dream of Bourbaki, to produce the modern equivalent of Euclid's *Elements*, had receded.

In his later years, he said, he had come to dwell more and more in history, which had not only occupied his time but given him a kind of social life. An imaginary social life, that is: as he immersed himself in the writings and correspondence of the great mathematicians of the past, men like Pierre de Fermat and Leonhard Euler had become "personal friends," he said. Their companionship had brought him happiness in his old age.

"Will such a statement edify and enlighten the present audience?" he asked at the close. "I am inclined to doubt it, but at my age, I fear it is the best I can do."

The mathematician walks off the stage.

The Weil conjectures were invoked by Barry Mazur, in a 2005 dialogue published in a humanities journal, to illustrate a larger point he wanted to make about mathematics: "Sixty years ago, André Weil dreamt up a striking way of very tightly controlling and counting the number of solutions of systems of polynomial equations over finite fields," he said, referring to the conjectures. Weil did so by proposing that a method could be developed in number theory analogous to one in topology for counting intersections of geometric subspaces. And this, said Mazur, goes to show that no mathematics, not even number theory, is divorced from our geometrical intuition. That is to say no mathematics is entirely cut off from the sensual.

Nor is it entirely cut off from the people who devised it. What Mazur's language suggests, beyond the broadest outline of a mathematical result, is conveyed in that phrase "very tightly controlling and counting"—hinting at Weil's mathematical personality, which was at once disciplined and visionary. I see him as a commander on the battlefield, a great strategist if at times a difficult person. "What makes his work unique in the mathematics of the twentieth century," wrote his friend and peer Jean-Pierre Serre after Weil's death, "is its prophetic aspect . . . combined with the utmost classical precision."

Maybe the dream of pristine writing, in which the writer is present but not present, masked behind the light of her brilliant transmission, is realized in these works I only dream of reading.

One August evening, a few months after sending the e-mail to the mathematician who never replied, I spot him at my local supermarket, headed for the bulk foods section. I push my cart in that direction but can't muster the will to go up to him, not just because of the unanswered e-mail but because a store employee is vacuuming the trays under the bulk food dispensers with a very noisy machine, also the mathematician is wearing earbuds, so I can only imagine an interaction with him as a desperate exchange of hand gestures. I hesitate, realize he's no longer among the bulk bins, try to turn around, but a wheel of my cart gets stuck on the vacuum cord, and by the time I work it free I figure him for lost.

I give up—for half a minute or so—and then I think, No, dammit, I'm going to find that guy and talk to him. I begin scanning the aisles, one by one, and finally corner the poor man by the tortillas. If there's one thing I had plenty of chances to practice during years of work as a journalist, it's approaching somebody I feel quite shy about approaching. Yet I'm not sure whether repetition has made it any easier. My smile feels like a sticker slapped onto my face.

Yes, he says, you sent me an e-mail. I'm sorry that I never replied—

Oh that's okay, I say. E-mails! Maybe there's someone else in your department that I could to talk to—

I'm probably the one, he says, slowly, not with any evident enthusiasm.

In other words, it's pretty awkward. What do I even want from him? I'm about to leave town for a couple of weeks, I tell him, and we agree, tentatively, that we'll have coffee once I'm back.

A very broad, bird's-eye overview of the development of twentieth-century mathematics (which I'm mostly borrowing from a lecture given by the mathematician Michael Atiyah) goes like this: In the first half of the century, the prevailing concern was to define and formalize things, as in the Bourbaki effort, and to pursue work specific to the different subject areas within mathematics. Then in the second half, mathematicians looked for ways to tie together those different areas, to transfer techniques from one to the other, and math became more global, even as it expanded so much that it became impossible for anyone to fully comprehend all that was happening in the field.

So one can faithfully call André Weil a pivotal figure in twentieth-century mathematics. He helped to found Bourbaki in his early career, emblematic of that era's determination to ground and make rigorous the subfields of math, and not long after that he began to envision connections between mathematical land masses, which his successors would build out.

It's very hard to conceive of how mathematicians from a long time ago thought about things, Atiyah said in that same lecture, because subsequent discoveries have become so ingrained. "In fact if you make a really important discovery in mathematics you will get omitted altogether!" said Atiyah in that same lecture. "You simply get absorbed into the background."

Another reckless dash through twentieth-century math might emphasize that as time went on (at least in some camps), identity became less important than relatedness. Theories were developed in which knowing relationships among mathematical objects matters more than knowing about the mathematical objects themselves. Or you could even say that knowing an object itself is the same thing as knowing its relations to other objects of the same kind.

A mathematics increasingly globalized, increasingly concerned with links from one thing to another, increasingly aided by computers, a highly connected world in which it's still very difficult to see much more than your own small part—remote as the world of math is, this all begins to seem familiar. Like that forest of links in which I keep losing my way.

But there's a second part to my supermarket chat with the mathematician, in which I learn that he has a son the same age as my son, and we take refuge in that, as parent-strangers will do. His son (go figure) likes math too. We compare notes about math-related books for kids, and somehow that makes it seem more possible that we will actually at some point get together and talk about the Weil conjectures.

"I can remember Bertrand Russell telling me of a horrible dream," wrote Godfrey Harold Hardy in his book *A Mathematician's Apology*. "He was in the top floor of the University Library, about A.D. 2100. A library assistant was going round the shelves carrying an enormous bucket, taking down books, glancing at them, restoring them to the shelves or dumping them into the bucket. At last he came to three large volumes which Russell could recognize as the last surviving copy of [Russell's own] *Principia Mathematica*. He took down one of the volumes, turned over a few pages, seemed puzzled for a moment by the curious symbolism, closed the volume, balanced it in his hand and hesitated . . ."

The mathematician is required to disappear.

Goro Shimura saw André Weil for the last time on a December afternoon in 1996, at the Institute for Advanced Study. Shimura, by then a professor at Princeton, had tried to speak to Weil by phone the day before, but Weil couldn't hear very well, and he'd asked Shimura to meet him at the institute instead. Come tomorrow, he said, otherwise I won't remember.

It had been drizzling endlessly, Shimura would later write. Forty-one years had passed since they first met in Tokyo. When he arrived at the institute, Shimura discovered that Weil had forgotten his hearing aid, and they drove to his house for it. Yet even with the device Weil couldn't hear well, and at lunch their conversation was halting and uncomfortable.

Shimura was working on a problem; he'd had a new idea about the Siegel mass formula, a topic that Weil had once studied, but when Shimura asked about the history of the subject, all Weil could say was, "I don't remember." He kept saying the same thing over and over: That was a long time ago, I don't remember.

Shimura asked whether he was still writing history.

I cannot write anymore, he said.

They left the dining hall and walked through the drizzle to Shimura's car.

You are certainly disappointed, but I am disappointed too, Weil said, with myself.

And then he said again, I cannot write anymore.

This episode appears in a reminiscence that Shimura wrote after Weil died, for the *Bulletin of the American Mathematical Society*. Like his earlier "Yutaka Taniyama and His Time," it's a beautiful tribute, and just based on those two essays I'm ready to pronounce Shimura the great elegist of twentieth-century mathematics.

Is that what I am writing, I wonder, some sort of elegy for math, or for my own entanglement with math? At times it feels that way, but I don't think that's what this is. As it turns out, one stilted encounter in the supermarket is enough to send me back to the algebra lectures, which for no good reason I still want to finish. And so it's on to ring theory, which is of course nothing I need to remember, nothing I need to know.

A mathematician might dream of an afterlife in which all is revealed, mathematical structures extending as far and wide and high as the eye can see, their nature and relations made transparent, and look, there are Galois and Archimedes strolling by, there Germain and Taniyama deep in conversation. And in that paradise, maybe, Simone and André would be reunited; they would find each other and argue in ancient languages all through the never-ending day.

Their earthly afterlives are, in a sense, opposites: the image of Simone the person weighs upon her writings, inseparable from them, even overtaking them, while André grows ever more attenuated. His person will vanish—be absorbed into the background—and leave behind just a name, a quartet of letters attached to other symbols, theorems and conjectures.

Where are numbers? my son asks. And where, for that matter, are all the unknown theorems, all the hidden proofs, all the math not yet discovered?

Thirty-five hundred years ago, a Babylonian presses a wedge into clay. One, two, three times. Then leaves a gap. Then presses again.

ACKNOWLEDGMENTS

Writing this book was an unexpected journey down a rabbit hole, and I've been extremely grateful for the help and support I received along the way. Early reads and encouragement from Cecily Parks, Louisa Hall, and Andrew Bujalski were crucial. Amy Williams's enthusiasm was heartening and infectious. I am also indebted to the following people who read drafts of the book and offered insightful comments: Ric Ancel, David Ben-Zvi, Pam Colloff, Lauren Meyers, Dominic Smith, and Kirk Walsh. And to Jessica Halonen for her wonderful art.

I count myself very fortunate to have Emily Bell as an editor, and I'd also like to thank everyone else at FSG, in particular Jackson Howard, Lottchen Shivers, Karla Eoff, Scott Auerbach, and Thomas Colligan.

Thanks also to the Ucross Foundation, and to Sharon Dynak and Tracey Kikut and Cindy Brooks in particular, for the gifts of time and space and sustenance.

Many books and articles informed this project, and of those I'd like to acknowledge *The Apprenticeship of a Mathematician*, by André Weil; Simone Weil's essays and letters; *At Home with André and Simone Weil*, by Sylvie Weil; *Simone Weil*, by Simone Pétrement; *Remarkable Mathematicians*, by Ioan James; *Bourbaki: A Secret Society of Mathematicians*, by Maurice Mashaal; *Mathematical Thought from Ancient to Modern Times*, by Morris Kline; and *Mathematics Without Apologies*, by Michael Harris. Excerpts from

André's 1940 prison letter to his sister are from Martin Krieger's translation.

Though I'm not able to name them all, I am also thankful for every schoolteacher, professor, teaching assistant, and fellow student who taught me math over the years.

A Note About the Author

Karen Olsson is the author of the novels *Waterloo* and *All the Houses*. She has written for *The New York Times Magazine*, *Slate*, *Bookforum*, and *Texas Monthly*, among other publications, and is a former editor of the *Texas Observer*. She graduated from Harvard University with a degree in mathematics and lives in Austin, Texas, with her family.